MATHEMATICS AND THE
SEARCH FOR KNOWLEDGE

Mathematics and the Search for Knowledge

Morris Kline

PROFESSOR EMERITUS
NEW YORK UNIVERSITY

OXFORD UNIVERSITY PRESS
New York   Oxford

Oxford University Press
Oxford   New York   Toronto
Delhi   Bombay   Calcutta   Madras   Karachi
Petaling Jaya   Singapore   Hong Kong   Tokyo
Nairobi   Dar es Salaam   Cape Town
Melbourne   Auckland

and associated companies in
Beirut   Berlin   Ibadan   Nicosia

Library of Congress Cataloging in Publication Data
Kline, Morris, 1908–
   Mathematics and the search for knowledge.
   Bibliography: p
   Includes index.
   1. Mathematics—1961–
2. Perception.
3. Knowledge, Theory of.
I. Title.
QA99.K55   1985   510   84-14809
ISBN 0-19-503533-X
ISBN 0-19-504230-1 (pbk.)

Printing (last digit): 9 8 7

Printed in the United States of America

Preface

How do we acquire knowledge about our physical world? All of us are obliged to rely on our sense perceptions—hearing, sight, touch, taste, and smell—to perform our daily tasks and enjoy some pleasures. These perceptions tell us a good deal about our physical world, yet in general they are crude. As Descartes put it, perhaps too strongly, sense perception is sense deception. It is true that modern instruments such as telescopes extend our perceptions immensely, yet such instruments have limited applicability.

Major phenomena of our physical world are not perceived at all by the senses. They do not tell us that the Earth is rotating on its axis and revolving around the sun. They do not tell us the nature of the force that keeps all the planets revolving around the sun. Nor do they tell us anything about electromagnetic waves that enable us to receive radio and television programs originating hundreds and even thousands of miles away.

This book will not be much concerned with what one might characterize as mundane applications of mathematics such as finding the precise height of a fifty-story building. The reader will be made aware of the limitations of sense perceptions, but our chief concern will be to describe what is known about the realities of our physical world *only* through the medium of mathematics. Rather than presenting much of the actual mathematics, I shall describe what mathematics reveals about major phenomena in our modern world. Of course, experience and experimentation play a role in our investigation of nature; however, as will become evident, these measures are in many areas minor.

In the seventeenth century, Blaise Pascal bemoaned human helplessness. Yet today a tremendously powerful weapon of our own creation—namely, mathematics—has given us knowledge and mastery of major areas of our physical world. In his address in 1900 at the International Congress of Mathematicians, David Hilbert, the foremost mathematician of our era, said: "Mathematics is the foundation of all exact knowledge of natural phenomena." One can justifiably add that, for many vital phenomena, mathematics provides the only knowledge we have. In fact, some sciences are made up solely of a collection of mathematical theories adorned with a few physical facts.

Contrary to the impression students acquire in school, mathematics is not just a series of techniques. Mathematics tells us what we have never known or even suspected about notable phenomena and in some instances even contradicts perception. It is the essence of our knowledge of the physical world. It not only transcends perception but outclasses it.

ACKNOWLEDGMENT

I am grateful to members of the Oxford University Press for their scrupulous attention to the preparation of this book. I also wish to thank my wife Helen and Miss Marilyn Manewitz for their careful reading and typing of the manuscript.

Brooklyn, New York   M.K.
March 1985

Contents

MATHEMATICS AND THE
SEARCH FOR KNOWLEDGE

Historical Overview: Is There an External World?

A philosopher is someone who knows something about knowing which nobody else knows so well.   Descartes

There is nothing so absurd that it has not been said by philosophers.   Cicero

Is not the whole of philosophy like writing in honey? It looks wonderful at first sight. But when you look again it is all gone. Only the smear is left.   Einstein

Is there a physical world independent of humanity? Are there mountains, trees, land, sea, and sky, all of which exist whether or not human beings are there to perceive these objects? The question seems silly. Of course there is. Do we not observe this world constantly? Do not our senses give us continuous evidence of the existence of this world? However, thoughtful people are not above questioning the obvious even if only to confirm it.

Let us begin by turning to the philosophers, the lovers of wisdom, who have for centuries pondered all matters concerning humanity and our world but who like unrequited lovers often felt rejected. Some of the greatest philosophers have considered this very subject of the existence of an external world. Some have denied it; others have admitted its existence but have seriously doubted how much we can know about the external world and how reliable our knowledge is. Although Bertrand Russell, himself an outstanding philosopher, said in *Our Knowledge of the External World,* "Philosophy, from the earliest times, has made greater claims and achieved fewer results than any other branch of learning," we should at least examine what some of the philosophers

3

have had to say. We shall be concerned primarily with those who have seriously questioned our knowledge of the external world.

The first of the ancient Greek philosophers to deal with this question was Heraclitus, who lived about 500 B.C. Heraclitus did not deny that there is an external world but maintained that everything in the world is constantly changing. As he put it, one cannot step twice into the same river, so whatever facts we may think we have gathered about the physical world no longer obtain in the very next instant.

Epicurus (341–270 B.C.), in contrast, held to the fundamental principle that our senses are the infallible guide to truth. They tell us that matter exists, that motion occurs, and that the ultimate realities are bodies composed of atoms existing in a void. They have existed forever and cannot be destroyed. They are indivisible and unchangeable.

Plato (427–347 B.C.), the most influential philosopher of all time, was also interested in this problem. He admitted the existence of an external world but came to the conclusion that the world perceived by the senses is motley, protean, ever-changing, and unreliable. The true world is the world of ideas, which is unchangeable and incorruptible. Yet this world of ideas is not accessible to the senses but only to the mind. Observations are useless. Thus, in the *Republic,* Plato states clearly that the reality behind appearances, what is inherently true of them, is mathematical; to understand reality is to elicit it from appearances, not to impose it on them. Mathematics is the foundation of true being, the "eternally real." In emphasizing the importance of mathematics Plato conceived of it as a part of a more general system of abstract, nonmaterial, ideal *Ideas.* These are models of perfection that everything in the universe—the material, the ethical, and the aesthetic—strives to attain. Plato says in the *Republic:*

> But if anyone tries to learn about things of the sense, whether gaping up or blinking down, I would never say that he really learns—for nothing of the kind admits of true knowledge—nor would I say that his soul looks up, but down, even though he study floating on his back on sea or land.

Plutarch relates in his "Life of Marcellus" that Eudoxus and Archytas, famous contemporaries of Plato, used physical arguments to "prove" mathematical results. However, Plato indignantly denounced such proofs as a corruption of geometry; they utilized sensuous facts in place of pure reasoning.

Plato's attitude toward astronomy illustrates his position on the knowledge that we should seek. This science, he said, is not concerned with the movements of the visible heavenly bodies. The arrangement of the stars in the heavens and their apparent movements are indeed wonderful and beautiful to behold, but mere observations and explanations of the motions fall far short of true astronomy. Before we can attain to this true science we "must leave the heavens alone," for true astronomy deals with the laws of motion of true stars in a mathematical heaven of which the visible heaven is but an imperfect expression. Through Socrates, in words now famous, Plato tells us the astronomer's true concerns:

> These sparks that paint the sky, since they are decorations on a visible surface, we must regard, to be sure, as the fairest and most exact of material things but we must recognize that they fall far short of the truth, the movements, namely, of real speed and real slowness in true number and in all true figures.... These can be apprehended only by reason and thought, but not by sight. [Therefore] we must use the blazonry of the heavens [merely] as patterns to aid in the study of those realities, if we are to have a part in the true science of astronomy.

This conception of astronomy staggers the modern mind, and scholars have not hesitated to indict Plato's downgrading of sensory experience as a grave disservice to the advancement of science. We should recognize, however, that the role here assigned to the astronomer parallels precisely the course followed successfully by the geometrician, who studies mental idealizations of, for example, triangles rather than particular triangular objects. In Plato's time observational astronomy had been pursued virtually to the limits then attainable, and he may have thought that further progress now awaited some hard thinking and theorizing on the assembled information.

Plato's conception of abstract ideals did unfortunately retard progress in experimental science for centuries. It implied that true knowledge could be acquired only through philosophical contemplation of abstract ideas and not through observation of the accidental and imperfect things in the real world.

However, there were and are philosophers who accepted the existence of an external world and believed that we can acquire sound knowledge from our sensations. Aristotle, in opposition to Plato, not only affirmed the existence of a world external to humanity but also maintained that our ideas about the world are obtained by abstracting from it ideas common to various classes of material objects we perceive such as triangles, spheres, foliage, and mountains. He criticized Plato's otherworldliness and his reduction of science to mathematics. Aristotle, physicist in the literal sense of the word, believed in material things as the primary substance and source of reality. Physics, and science generally, must study the physical world and obtain truths from it. Thus, genuine knowledge is obtained from sense experience, by intuition and by abstraction. These abstractions have no existence independent of human minds.

To arrive at truths, Aristotle used what he called universals—general qualities that are abstracted from real things. In his words, we must "start with things which are knowable and observable to us and proceed toward those things which are clearer and more knowable by nature." He took the obvious sensuous qualities of objects, hypostatized them, and elevated them to independent mental concepts. Specifically, above the central Earth, which includes the whole volume of water, comes the region occupied by air; higher still, extending as far as the moon, is a substance that we call fire, although in reality it is a mixture of fire and air. They owe their existence, says Aristotle, to "four principles": hot, cold, dry, and moist (see Chapters V and X). These can combine in pairs in six ways, but two of the combinations—hot with cold, dry with moist—are inherently impossible; the remaining four pairs generate the four elements. Thus, earth is cold and dry; water, cold and moist; air, hot and moist; and fire, hot and dry. The elements are not eternal; on the contrary, matter is continually passing from one form to another. All the universe, from Earth out as far as the moon, is a region undergoing constant change, corruption, mortality, and decay, as the phenomena of climate and geology vividly attest.

Although their influence is unmistakable, one might be inclined to dismiss the views of Greek philosophers, because while their culture emphasized mathematics, they lived in what may justifiably be called a prescientific world. They did not experiment much and, on the whole, lived apart from the world of science as we know it today.

During the Middle Ages concern about the external world was unimportant; theology was the supreme concern. Not until the Renaissance did philosophers turn with heightened interest to the physical world. In Western Europe especially, we can see the beginning of modern philosophy and with it, a new interest in science.

René Descartes (1596–1650) is the founder of modern philosophy. His *Discourse on the Method of Rightly Conducting the Reason and Seeking Truth in the Sciences* (1637), which contained three appendices, "Geometry," "Dioptric," and "Meteors," is a classic. Although Descartes thought that his philosophical and scientific doctrines subverted Aristotelianism and scholasticism, he was at heart a scholastic or Aristotelian. He followed in the footsteps of Aristotle and drew from

his own mind propositions about the nature of being and reality
haps for this very reason his writings influenced the seventeenth
tury more extensively than the researches of those who had begun to
draw truths from observation and experimentation—sources that were
wholly at variance with the traditional ones.

Seeing that there was the logical possibility that all his beliefs were
false, Descartes sought a solid base on which he might build an edifice
of truth. He found but one fact he could be sure of—*Cogito, ergo sum*
(I think, therefore I am). Because he recognized that he was finite and
imperfect, he reasoned that this very sense of his limitations implied
that there had to be an infinite and perfect being against whom he measured himself. This being, God, had to exist because He would not be
perfect if he lacked the essential attribute of existence. To Descartes
this result, God's existence, was more important for science than for
theology, for it afforded the possibility of solving the central problem
of the existence of an objective world.

Because all our knowledge of a world external to our minds comes
to us through sense impressions, the question arises whether there
exists anything other than just these, or whether objective reality is an
illusion. To this question Descartes answered that God, being perfect,
would not be a deceiver; He would not make us believe in the existence
of a material universe if it were not real.

This objective reality can be grasped primarily through the physical attribute of extension. This is innate in the very notion of matter,
which itself is not derived from the senses. Therefore, no knowledge
of the material world is, save in an indirect manner, derived from the
senses. Descartes was also able to classify his observations of material
objects into primary and secondary qualities. Thus, color is secondary
because it is perceived only by the senses, whereas extension and
motion are primary.

To Descartes, the entire physical universe is a great machine operating according to laws that may be discovered by human reason, particularly mathematical reasoning. He deprecated experimentation,
although he did experiment in biology.

Responding directly to the knowledge that was being acquired in
mathematics and science, the philosopher Thomas Hobbes (1588–
1679) affirmed in his *Leviathan* (1651) that external to us there is only
matter in motion. External bodies press against our sense organs and
by purely mechanical processes produce sensations in our brains. All
knowledge is derived from these sensations, which then become
images in the brain. When a train of images arrives, it recalls others

already received—as, for example, the image of an apple recalls that of a tree. Thought is the organization of chains of images. Specifically, names are attached to bodies and properties of bodies as they appear in images, and thought consists in connecting these names by assertions and in seeking the relations that necessarily hold among these assertions.

Hobbes, in his book *Human Nature* (1650), says that ideas are images or memories of what is received through the senses. There are no innate ideas or ideals; no universals or abstract ideas. Triangle means merely the idea (image) of all triangles perceived. All substance that gives rise to ideas is material or corporeal. In fact, the mind too is substance. Language (for example, the language of science and mathematics) consists only of symbols or names for experiences. All knowledge is but remembrance, and the mind works with words that are but names for things. True and false are attributes of names, not of things. That humans are living creatures is true because whatever is called human is also called a living creature.

Knowledge is obtained when regularities are discovered by the brain as it organizes and relates the assertions about physical objects. And mathematical activity produces just such regularities. Hence the mathematical activity of the brain produces genuine knowledge of the physical world, and mathematical knowledge *is* truth. In fact, reality is accessible to us only in the form of mathematics.

So strongly did Hobbes defend the exclusive right of mathematics to truth that even the mathematicians objected. In a letter to a leading physicist of the age, Christian Huygens, the mathematician John Wallis wrote of Hobbes:

> Our Leviathan is furiously attacking and destroying our Universities (and not only ours but all) and especially ministers and the clergy and all religion, as though the Christian world had no sound knowledge, none that was not ridiculous either in philosophy or religion, and as though men could not understand religion if they did not understand philosophy, nor philosophy unless they knew mathematics.

The emphasis placed by Hobbes on the purely physical origin of sensation and on the limited action of the brain in reasoning shocked many philosophers to whom the mind was more than a mass of matter acting mechanically. In his *Essay Concerning Human Understanding,* published in 1690, John Locke (1632–1704) began somewhat as Hobbes did, but unlike Descartes, by asserting that there are no innate ideas in humans; men are born with minds as empty as blank tablets.

Experience, through the media of the sense organs, writes on those tablets and produces simple ideas. Some simple ideas are exact resemblances of qualities actually inhering in bodies. These qualities, which he called primary, are exemplified by solidity, extension, figure (shape), motion or rest, and number. Such properties exist whether or not anyone perceives them. Other ideas that arise from sensations are the effects of the real properties of objects on the mind, but these ideas do not correspond to actual properties. Among such secondary qualities are color, taste, smell, and sound.

Locke's aim in the *Essay* was to discover the limit or boundary between the knowable and the unknowable, the "horizon . . . which sets the bounds between the enlightened and dark parts of things." By so doing, he would refute both the skeptics, who "question everything, and disclaim all knowledge, because some things are not to be understood," and at the opposite extreme those overconfident reasoners who presume that the whole vast ocean of being is "the natural and undoubted possession of our understanding, wherein there was nothing exempt from its decisions, or that escaped its comprehension." More positively and constructively, his purpose was to establish the grounds of knowledge and opinion, and the measures by which truth might be attained or approximated in all things that the human understanding has the capacity to comprehend.

The plan or design of the *Essay,* as Locke explained in the introduction, was to "inquire into the origin, certainty, and extent of *human knowledge;* together with the grounds and degrees of *belief, opinion,* and *assent.*" Following a "historical, plain method," he gave an account of the origin of our ideas, then showed what knowledge the understanding had through those ideas, and finally inquired into the nature and grounds of faith or opinion.

Although the mind cannot invent or frame any simple idea, it does have the power to reflect on, compare, and unite simple ideas and thus form complex ideas. Here Locke departed from Hobbes. In addition, he said, the mind does not know reality itself, but only ideas of reality, and it works with these. Knowledge concerns the connection of ideas such as their agreement or inconsistency. Truth consists in knowledge that conforms to the reality of things.

Basic mathematical ideas are constructed by the mind, although they are ultimately traceable to experience; however, some ideas are not traceable to real entities. These latter, more abstract mathematical ideas are constructed from the former by repeating, combining, and arranging them in various ways. Perception, thinking, doubting, believing, reasoning, willing, and knowing generate these latter abstract ideas. Thus one obtains the idea of a perfect circle. There is, then, an internal sense that produces the abstract ideas. Mathematical knowledge is universal, absolute, certain, and significant. This knowledge is real, even though it consists of ideas.

Demonstration connects these ideas and thereby establishes truths. Locke preferred mathematical knowledge because he felt that the ideas with which it deals were the clearest and most reliable. Furthermore, mathematics relates ideas by exhibiting necessary connections among them, and the mind understands such connections best. Not only did Locke prefer the mathematical knowledge of the physical world produced by science, he even rejected the direct physical knowledge, arguing that many facts about the structure of matter are simply not clear, such as the physical forces by which objects attract or repel each other. Moreover, because we can never know the real substance of the external world but only ideas produced by sensations, physical knowledge can hardly be satisfactory. He was convinced, nevertheless, that the physical world possessing the properties described by mathematics does exist, as do God and we ourselves.

In general, Locke's theory of knowledge, though somewhat ambiguous, may be described as intuitional. In his system, truth inheres solely in propositions, and the way to advance both knowledge and right judgment is by comparing propositions, either directly or through intermediary ideas, to determine their agreement or disagreement with each other. Knowledge is possible wherever this agreement or disagreement is immediately and certainly perceptible.

Even in demonstrative reasoning, in which the agreement or disagreement is not directly perceived but must be established by the intervention of other ideas, each step of the argument must be intuitively clear and certain. Another source of knowledge is sensation, by which we *intuit* the existence of particular external things, when they are present to our senses.

From the first of these sources, direct intuition, we have certain knowledge of our own existence, because in "every act of sensation, reasoning, or thinking, we are conscious to ourselves of our own being; and, in this matter, come not short of the highest degree of certainty." The mathematical relations of geometry and algebra, the principles of abstract morality, and the existence of God may all be proved by reasoned demonstration, while the existence of external things when actually present to the senses is, of course, known by sensation. These are all fundamental truths of the most vital importance for our existence and well-being, both here and hereafter, but obviously they do not take us far into the vast ocean of being and life.

Locke, like Descartes, discarded all secondary qualities. Nature is a dull affair, soundless, senseless, scentless and colorless and restricted to motion of meaningless material. Locke's influence on popular thought was enormous, and his philosophy pervaded the eighteenth century much as Descartes's did in the seventeenth.

In their theories of knowledge Hobbes and, to a lesser extent, Locke put primary emphasis on the existence of a world of matter external to human beings. While all knowledge stemmed from this source, the surest truths about this world finally obtained by the mind, or brain, were the laws of mathematics. Bishop George Berkeley (1685–1753), famous as a philosopher as well as a churchman, recognized in this emphasis on matter and mathematics the threat to religion and to concepts such as God and the soul. With ingenious and trenchant arguments he proceeded to attack both Hobbes and Locke and to offer his own theory of knowledge.

He is most radical in denying the existence of an external world. Basically, his argument is that all sensations are subjective and thus dependent on the observer and his or her point of view. He accounts for the seeming permanence of many conscious perceptions (for example, that a tree appears unchanged when seen on two successive occasions) by the assertion that they persist in the mind of God.

In his chief philosophical work, *A Treatise Concerning the Principles of Human Knowledge* (1708), where he considered the leading causes of error and difficulty in the sciences and the grounds of skepticism, atheism, and irreligion, Berkeley made a frontal assault. Both Hobbes and Locke had maintained that all we know are ideas produced by the action of external, material objects on our minds. Berkeley granted the sensations or sense impressions and the ideas derived from them, but he challenged the belief that they are caused by material objects external to the perceiving mind. Because we perceive only the sensations and the ideas, there is no reason to believe that anything is external to ourselves. In response to Locke's argument that our ideas of the primary qualities of material objects are exact copies, Berkeley retorted that an idea can be like nothing but an idea:

> When we do our utmost to conceive the existence of external bodies, we are all the while contemplating our own ideas. But the mind, *taking no*

notice of itself, is deluded into thinking it can and does conceive bodies existing with aught of or without the mind.

All our knowledge is in the mind.

Berkeley strengthened his position with an argument unintentionally supplied by Locke when he had distinguished ideas of primary qualities from those of secondary qualities. The former corresponded to real properties whereas the latter existed only in the mind. Berkeley asked: Can anyone conceive of the extension and motion of a body without including other sensible qualities, such as color? Extension, figure, and motion per se are inconceivable. If, therefore, the secondary qualities exist only in the mind, so do the primary ones.

In brief, Berkeley argued that because we know only sensations and ideas formed by these sensations but do not know external objects themselves, there is no need to assume an external world at all. That world does not exist any more than do the stars one sees after receiving a blow on the head. An external world of matter is a meaningless and incomprehensible inference. If there were external bodies, we should never be able to know it; and if there were not, then we should have the same reasons as now to think that there were such bodies. Mind and sensations are the only realities. Thus Berkeley disposed of matter.

However, Berkeley had yet to reckon with mathematics. How was it that the mind was able to obtain laws that not only described but predicted the course of the supposed external world? What could he do to counter the strongly established eighteenth-century belief in the truths about an external world proffered by mathematics?

He proceeded to demolish mathematics and was shrewd enough to attack it at its weakest point. The fundamental concept of the calculus is the instantaneous rate of change of a function, but this concept was not clearly understood and therefore not well presented by either Newton or Leibniz. Hence Berkeley was able in his day to attack it with justification and conviction. In *The Analyst* of 1734, addressed to an infidel mathematician (Edmund Halley), he did not mince words. Instantaneous rates of change he condemned as "neither finite quantities nor quantities infinitely small, nor yet nothing." These rates of change were but "the ghosts of departed quantities. Certainly . . . he who can digest a second or third fluxion [Newton's technical name for instantaneous rate of change] need not, methinks, be squeamish about any point in Divinity." That the calculus proved useful nevertheless, Berkeley accounted for on the grounds that somewhere errors were compensating for each other. Although Berkeley had made a criticism of the calculus that was warranted at that time, he had not actually disposed of all the truths mathematics had produced about the physical world. Nevertheless, having given his opponents something to think about, he rested his case against mathematics at this point. Berkeley summed up his philosophy in this way:

> All the choir of heavens and furniture of earth, in a word all those bodies which compose the mighty frame of the world, have not any substance without the mind . . . So long as they are not actually perceived by me, or do not exist in my mind, or that of any other created spirit, they must either have no existence at all, or else subsist in the mind of some Eternal Spirit.

Even Berkeley himself was not above an occasional sortie into the very physical world whose existence he denied. His last work, entitled *Siris: A Chain of Philosophical Reflections Concerning the Virtues of Tar-Water,* recommended the drinking of water in which tar had been soaked as a cure for smallpox, consumption, gout, pleurisy, asthma, indigestion, and many other diseases. Such occasional missteps must not be held against Berkeley. The reader who consults his delightful *Dialogues of Hylas and Philonous* will find an extremely able and entertaining defense of his philosophy.

Berkeley's extreme views on matter and mind led to the pun: "What is matter? Never mind. What is mind? Never matter." At any rate, by depriving materialism of its matter, Berkeley believed he had disposed of the physical world.

It would seem that Berkeley's philosophy was about as radical as thought can be on the subject of humanity's relation to the physical world. However, the skeptic Scot, David Hume (1711–1776), thought Berkeley had not gone far enough. While Berkeley accepted a thinking mind in which sensations and ideas existed, Hume even denied mind. In his *Treatise of Human Nature* (1739–1740), he maintained that we know neither mind nor matter. Both are fictions, neither of which we perceive. We perceive impressions (sensations) and ideas, such as images, memories, and thoughts, all three of which are but faint effects of impressions. There are, it is true, both simple and complex impressions and ideas, but the latter are merely combinations of simple ones. Hence it can be asserted that the mind is identical with our collection of impressions and ideas. It is but a convenient term for this collection.

As for matter, Hume agreed with Berkeley. Who guarantees to us that there is a permanently existing world of solid objects? All we *know* are our own sensory impressions of such a world. By association of ideas through resemblance and contiguity in order or position, the memory orders the mental world of ideas much as gravitation presumes to order the physical world. Space and time are only a manner and order in which ideas occur to us. Neither space nor time are objective realities. We are deluded by the force and firmness of our ideas into believing in such realities.

The existence of an external world with fixed properties is really an unwarranted inference. There is no evidence that anything exists beyond impressions and ideas that belong to nothing and represent nothing. Hence there can be no scientific laws concerning a permanent, objective physical world; such laws signify merely convenient summaries of impressions. Moreover, we have no way of knowing that the sequences we have observed will recur. In fact, we ourselves are but isolated collections of perceptions, that is, impressions and ideas. We exist only as such. Any attempt on our part to perceive ourselves reaches only a perception. All the other people and the supposed external world are just perceptions to any one person, and there is no assurance that they exist.

Only one obstacle stood in the way of Hume's thoroughgoing skepticism, namely, the existence of the generally acknowledged truths of pure mathematics itself. He could not demolish these, so he proceeded to deflate their value. The theorems of pure mathematics, he asserted, were no more than redundant statements, needless repetitions of the same facts in different ways. That two times two equals four is no new fact. Actually, two times two is but another way of saying or writing four. Hence this and other statements in arithmetic are mere tautologies. As for the theorems of geometry, they are but repetitions in more elaborate form of the axioms, which in turn have as much meaning as two times two equals four.

Specifically, Hume in his *Treatise of Human Nature* was skeptical about the power of reason as a tool for rational explanation.

> No object ever discloses by the qualities which appear to the senses, either the causes which produced it, or the effects which will arise from it; nor can our reason unassisted by experience ever draw any inference concerning real existence and matter of fact.

Experience may suggest a cause and effect, but this belief is never rational. A belief is rational only if its denial is logically inconsistent, but no belief arrived at through experience meets this demand. There is no real science about a permanent and objective world; science is purely empirical.

Hume's solution, then, of the general problem of how we obtain truths is that we cannot obtain them. Not the theorems of mathematics; nor the existence of God; nor the existence of an external world, causation, nature; nor miracles constituted truths. Thus Hume destroyed by reasoning what reasoning had established, while at the same time he emphasized the limitations of reason.

However, such a conclusion, such a denial of humanity's highest faculty, was revolting to most eighteenth-century thinkers. Mathematics and other manifestations of human reason had accomplished too much to be so easily cast aside. Immanuel Kant (1724–1804) actually expressed his revulsion for Hume's unwarranted extension of Locke's theory of knowledge: reason must be re-enthroned. It appeared indubitable to Kant that humanity possesses ideas and truths beyond mere amalgamations of sense experience.

Yet the outcome of Kant's cogitations, when carefully examined, was not much more comforting. In his *Prolegomena to Any Future Metaphysics* (1783), Kant wrote:

> We can say with confidence that certain pure a priori synthetical cognitions, pure mathematics and pure physics, are actual and given; for both contain propositions which are thoroughly recognized as absolutely certain . . . and yet as independent of experience.

In his *Critique of Pure Reason* (1781) Kant offered more reassuring words. He affirmed that all axioms and theorems of mathematics were truths. But why, Kant asked himself, was he willing to accept such truths? Surely experience itself did not vouchsafe it. The question could be answered if one could answer the larger question of how the very science of mathematics is possible.

In effect, Kant undertook an entirely new approach to the problem of how humanity obtains truths. His first step was to distinguish between two kinds of statements or judgments that give us knowledge. The first kind, which he called *analytical,* does not really contribute to knowledge. It is exemplified by the statement, *All bodies are extended.* This is merely an explicit statement of a property that bodies have by the very fact that they are bodies and says nothing new (although the statement may perhaps serve for emphasis). The second kind of knowledge, somehow obtained by the mind independently of experience, he called a priori knowledge.

According to Kant, truth cannot come from experience alone, for experience is a mélange of sensations, devoid of concepts and organization. Mere observations therefore will not furnish truths. Truths, if they exist, must be a priori judgments and, moreover, to be genuine knowledge they must be synthetic judgments; they must offer new knowledge.

Patent evidence was at hand in the body of mathematical knowledge. Almost all of the axioms and theorems of mathematics were to Kant a priori synthetic judgments. The statement that the straight line is the shortest distance between two points is certainly synthetic, because it combines two ideas—straightness and shortest distance—neither of which is implied by the other. Also it is a priori, in that experience with straight lines or even measurements could not ensure the invariable and universal truth Kant believed this statement to be. Hence to Kant there was no question that human beings do have a priori synthetic judgments, that is, genuine truths.

Kant probed still deeper. Why, he asked, was he willing to accept as a truth the statement that the straight line is the shortest distance between two points? How is it possible for the mind to know such truths? This question could be answered if we could answer the question of how mathematics is possible. The answer Kant gave is that our minds possess, independently of experience, the forms of space and time. Kant called these forms intuitions. They are pure a priori means of knowledge, not based on experience or thought. Space and time are therefore intuitions through which the mind necessarily views the physical world in order to organize and understand sensations. Since the very intuition of space has its origin in the mind, certain axioms about space are at once acceptable to the mind. Geometry then goes on to explore the logical implications of these axioms. The laws of space and time, laws of the mind, precede and make possible understanding of real phenomena. Kant said, "Our intellect does not draw its laws from nature but imposes its laws on nature."

We perceive, organize, and understand experience in accordance with these mental forms that are in the mind. Experience fits into them as dough into a mold. The mind imposes these modes on the received sense impressions and causes these sensations to fall into built-in patterns. Because the intuition of space has its origin in the mind, the mind automatically accepts certain properties of this space. Principles such as that the straight line is the shortest path between two points and that three points determine a plane, as well as the parallel axiom of Euclid, which Kant called a priori *synthetic* truths, are part of our mental equipment. The science of geometry merely explores the logical consequences of these principles. The very fact that the mind views experience in terms of the "spatial structure" of the mind means that experience will conform to the basic principles and the theorems.

Because Kant manufactured space from the cells of the human brain, he could see no reason not to make it Euclidean. His inability to conceive of another geometry convinced him that there could be no other. Thereby he guaranteed the truth of Euclidean geometry and at the same time the existence of a priori synthetic propositions. Thus the laws of Euclidean geometry were not inherent in the universe, nor was the universe so designed by God; they were humanity's mechanism for organizing and rationalizing sensations. As for God, Kant said that the nature of God fell outside of rational knowledge, but we should believe in Him. Kant's boldness in philosophy was surpassed by his rashness in geometry, for despite never having been more than ten miles from his home city of Koenigsberg in East Prussia, he could still determine the geometry of the world.

What about the mathematical laws of science? Because all experience is grasped in terms of the mental framework of space and time, mathematics must be applicable to all experience. In his *Metaphysical Elements of Natural Science* (1786), Kant accepted Newton's laws and their consequences as self-evident. He claimed to have demonstrated that Newton's first law of motion can be derived from pure reason and that this law is the only assumption under which nature is conceivable to human reason.

More generally, Kant argued that the world of science is a world of sense impressions arranged and controlled by the mind in accordance with innate categories such as space, time, cause and effect, and substance. The mind contains furniture into which the guests must fit. The sense impressions do originate in a real world, but unfortunately this world is unknowable. Actuality can be known only in terms of the subjective categories supplied by the perceiving mind. Hence there never would be another way to organize experience than by Euclidean geometry and Newtonian mechanics.

For Kant, as experience broadens and as new sciences are formed, the mind does not formulate new principles by generalizing from these new experiences; instead, unused compartments of the mind are called into use to interpret these new experiences. The mind's power of vision is illuminated by experience. This accounts for the relatively late recognition of some truths—for example, the laws of mechanics—compared with others known for many centuries.

Kant also said that we cannot hope to acquire indubitable knowledge from mere sensory acquaintance with objects. We shall never know real things in themselves. If we are capable of knowing anything with assurance, this must be a result of the processes that take place in our minds when they examine the data received from the external world.

Kant's philosophy, barely intimated here, glorified reason; however, he assigned to it the role of exploring not nature but the recesses of the human mind. Experience received due recognition as a necessary element in knowledge, because sensations from the external world supply the raw material that the mind organizes. Moreover, mathematics retained its place as the discloser of the necessary laws of the mind.

It is evident from this sketch of Kant's theory of knowledge that he made the existence of mathematical truths a central pillar of his philosophy. In particular, he relied on the truths of Euclidean geometry. Alas, the nineteenth-century creation, non-Euclidean geometry, demolished Kant's arguments.

Despite Kant's superlative philosophy and the regard with which his work was held, John Stuart Mill (1806–1873), the most celebrated English philosopher of the nineteenth century, returned to and somewhat altered Hume's position. Mill was a positivist, asserting that while knowledge comes primarily through the senses, it also includes the relations that the conscious mind formulates about the evidence of the senses. There is no way of proving that an external world exists, but there is also no proof that it does not exist.

By an external object we mean something that exists, whether it is thought of or not, that stays the same even if the sensations we get from it change, and is common to many observers in a way these sensations are not. To Mill, one's concept of the external world is made up only to a slight degree, at any moment, of actual sensations but to a large degree of possible sensations—not of what one is sensing but what one would sense upon moving or turning one's head. Matter is a permanent possibility of sensation. Memory plays a role in this type of knowledge.

Still, knowledge of the external world is known only by sensations. Such knowledge is imperfect, and we do not know its precise bounds and extent. The simple ideas we get from sensations are combined by our minds into a complex idea; however, this knowledge is only nominal, not essential. Inductive knowledge is never certain, only probable; yet it is the best we have for science and as a guide to our lives.

For Mill, our conclusions in mathematics, as in Euclidean geometry, are necessary only in the sense that they follow from the premises.

However, the premises—the axioms—are based on observations and are generalizations of experience. Arithmetic and algebra are also founded on experience. The expressions $2 + 2 = 3 + 1 = 4$ are psychological generalizations. Algebra is simply a more abstract extension of such generalizations.

So Mill found that induction is of central importance because it is the source of possible generalizations, such as the laws of nature. Cause is merely the antecedent of subsequent happenings. That every occurrence has a cause is derived from experience. It is a precise way of stating the principle of the uniformity of nature.

Thus nothing beyond experimental knowledge is either possible or necessary. Experience and psychology can explain all our knowledge and they are the basis of our belief in an external world. Mill is an empiricist, although he differs with Hume's skepticism. His ideas are close to and can even be said to initiate the empiricism and logical positivism of the twentieth century.

What conclusions can we draw from this survey of leading philosophers and their views concerning the existence of an external world and the reliability of our knowledge?* We take the position stated by Einstein:

> The belief in an external world independent of the percipient subject is the foundation of all science. But since our sense-perceptions inform us only indirectly of this world or Physical Reality it is only by speculation that it can become comprehensible to us.

Experience cannot prove reality. Experience is personal.

Although we take the position of the empiricists and undertake to see what we can learn about the external world, we should begin with the awareness of how reliable or unreliable our sense perceptions are. This we propose to do in Chapter I. Just what mathematics can do to correct what one might call illusions and especially to reveal totally unperceived physical phenomena will be our major concern.

---

*We shall say more on the views of modern philosophers who were influenced by the very creations we intend to discuss in later chapters.

I

The Failings of the Senses and Intuition

Sense perceptions are sense deceptions.   Descartes

Despite the denials of Berkeley, the qualifications of Hume, and the reservations of Heraclitus, Plato, Kant, and Mill concerning what we can know about the external world, physicists and mathematicians do believe that there is an external world. They would argue that even if all human beings were suddenly wiped out, the external or physical world would continue to exist. When a tree crashes to the ground in a forest, a sound is created even if no one is there to hear it. We have five senses—sight, hearing, touch, taste, and smell—and each of these constantly receives messages from this external world. Whether or not our sensations are reliable, we do receive them from some external source.

For practical reasons, such as remaining alive and possibly for improving life in the external world, we certainly want to know as much as possible about this world. We must distinguish land from sea. We must grow food, build shelters, and protect ourselves against wild beasts. Why then can we not rely on our senses to achieve these aims? Primitive civilizations have done so. But just as to the pure in heart the world is pure, so to the simpleminded the world is simple.

In attempting to improve our material way of life, we were forced to extend our knowledge of the external world. Thus we necessarily extended the uses of our senses to the utmost. Unfortunately for us, our senses are not only limited but also deceiving. Trusting solely in the senses can even lead to disaster. Let us note some of these limitations.

Of the five senses, the sense of sight is perhaps the most valuable.

Let us first test how much we can depend on this sense. We begin with a few simple examples. Through the years, many deceptive visual figures were deliberately conceived and constructed to show the limitations of the eye. In fact, physicists and astronomers of the nineteenth century took a lively interest in visual illusions because they were concerned that visual observations might prove to be unreliable. Wilhelm Wundt, an assistant to the famous physiologist, physician, and scientist Hermann von Helmholtz (1821–1894), constructed Figure 1. The illusion is simply that the vertical line looks longer than the horizontal line, which is of equal length. This illusion can be reversed. In Figure 2 the height and width seem equal, but the width is larger.

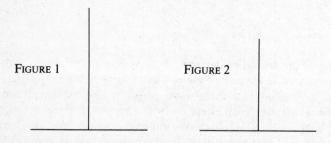

FIGURE 1   FIGURE 2

The illusion in Figure 3 was devised by Franz Müller-Lyer in 1899 and is also known as the Ernst Mach illusion. The two horizontal lines are actually the same length.

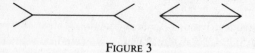

FIGURE 3

In Figure 4 the dot is at the middle of the horizontal line. Both illusions are created by angles.

FIGURE 4

In Figure 5 the length of the top horizontal line of the lower figure appears shorter than that of the top line of the upper figure. Incidentally, it is difficult to believe that the maximum horizontal width of the lower figure is as great as the maximum height of that figure.

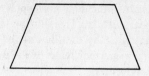

FIGURE 5

A striking illusion involving the influence of angles is shown in Figure 6. Here the diagonals *AB* and *AC* of the two parallelograms are of equal length, but the one on the right appears much shorter.

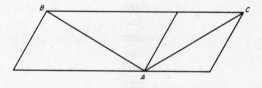

FIGURE 6

If oblique lines are extended across vertical ones, as in Figure 7, the illusion is very striking. The oblique line on the right if extended would meet the upper end of the oblique line on the left; however, the *apparent* point of intersection is somewhat lower. This well-known illusion is attributed to Johann P. Poggendorf (about 1860).

FIGURE 7

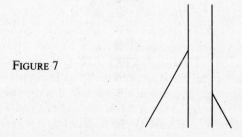

In Figure 8 the three horizontal lines are of equal length, although they appear unequal. This illusion is primarily a result of the sizes of the angles made by the lines at the ends. Within certain limits, the greater the angle the greater is the apparent elongation of the central horizontal portion.

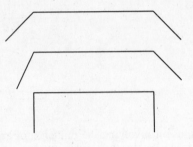

FIGURE 8

A striking illusion of contrast is shown in Figure 9. The central circles of the two figures are equal, although the one surrounded by the large circles appears much smaller than the one surrounded by the smaller circles.

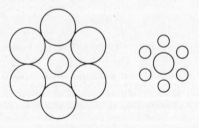

FIGURE 9

Another mechanism is believed to be at work in the Müller-Lyer illusion. In the left drawing of Figure 10 the horizontal lines at each end of the vertical line *A* are interpreted as the upper and the lower edges of two walls meeting at a corner. In this case the vertical line would be interpreted as being the foreground of a real-world scene. In the right drawing of Figure 10 the horizontal lines are again seen as wall edges, but in this case they appear to be converging on an inside corner. As a result the vertical line *B* is interpreted as being in the background. The judgment of size constancy then enlarges the perceived length of line *B* and diminishes line *A*.

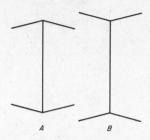

FIGURE 10

Johann Z. Zöllner was the first to describe the illusion illustrated in Figures 11 and 12. Zöllner accidentally noticed the illusion on a pattern designed for dress goods. The long parallel lines in Figure 11 appear to diverge, whereas in Figure 12 they appear to converge.

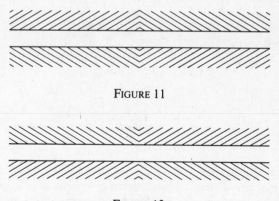

FIGURE 11

FIGURE 12

In the Hering illusion (Figure 13), published in 1861 by Ewald Hering, the straight horizontal lines acquire the illusion of curving in relation to the converging oblique lines.

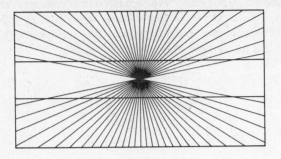

FIGURE 13

The unreliability of vision is demonstrated by still another example devised by Professor S. Tolansky. In Figure 14, commonly used in statistical studies, the baseline *CD* is as long as the height of the figure. Moreover, a viewer, when asked to draw a line across the curve that is half the width of the baseline, would almost surely choose the line *AB*. However, the correct choice is *XY*.

FIGURE 14

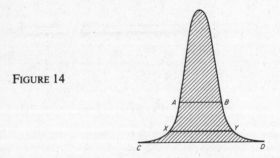

We are all familiar with an illusion that is deliberately and skillfully fashioned, namely, realistic painting. The intent is to present a three-dimensional scene on a flat or two-dimensional canvas. One of the great achievements of the Renaissance painters was to devise a mathematical scheme, known as linear perspective, which achieves the desired illusion.

Some simple examples of the illusion of linear perspective are part of our everyday experience. The principle involved in these examples and in the theory of linear perspective is that lines in the actual scene that recede directly from the observer must appear to come together at a distant point called the vanishing point. A simple example is furnished by the appearance of two parallel railroad tracks (Figure 15) that appear to meet at a distant point.

FIGURE 15

FIGURE 16

FIGURE 17

The influence of perspective is particularly apparent in Figure 16, where the usual perspective lines are drawn to suggest a scene. The tall boxes are of the same size and physical dimensions, but the farthest one actually appears much larger. Experience, expecting a diminution of size with increasing distance, actually causes the box on the right to appear larger than it really is.

We allow ourselves to be deceived and even enjoy the deception when we admire realistic painting. Such paintings must of necessity be two-dimensional, but if they are executed in accordance with the laws of linear mathematical perspective, we believe we are looking at a three-dimensional scene. Raphael's *School of Athens* (Figure 17) is a good example.

Of course, the mathematical system of linear perspective takes advantage of optical illusions. Objects or people that are intended to appear more distant than those in the foreground are drawn smaller, which is how the human eye would see them. Artists also take advantage of another optical effect, the loss of intensity or brightness of a distant object.

There are other visual illusions in our daily experiences. The sun and the moon seem larger when on the horizon than when overhead, because we unconsciously allow for our belief that they are closer when on the horizon. Precise measurement would of course show that their size remains constant.

If we measure the angle subtended at the eye by the diameter of the moon, we shall find it to be almost exactly one-half a degree. Because the whole semicircular vault of the heavens subtends 180 degrees, the angle subtended by the moon is a mere 1/360 of the vault of the heavens. The proportionate *area* occupied by the moon is the astonishingly small amount of only 1/100,000 of the heavenly vault, but if we consider how striking an object the full moon is, it is hard to appreciate how small a fraction of the sky area it occupies.

A number of other illusions involve what is known as the refraction or bending of light. All of us have noticed that a stick partially immersed in water seems bent and that the bending occurs at the surface of the water.

An aerial refraction phenomenon that has attracted attention since ancient times is the *mirage,* a phenomenon that comes about by the effect of variation in air density produced by the heat of the sun, combined with a total reflection effect. A simple mirage with which most of us are familiar occurs when one travels along a long, straight, flat road on a hot summer day. Well ahead the road appears to be covered with water, yet on traveling further one finds that the road is quite dry. Let us consider what causes this effect.

The effect only appears when the road surface is heated strongly by the sun. The air in contact with the road thus warms up, which lowers its density and causes it to rise continuously. It follows that the refraction of light is also less at the bottom than in the top layers of air. Let us imagine, as in Figure 18, that there is a succession of changing layers. Light passes through the layers and comes to our eyes from

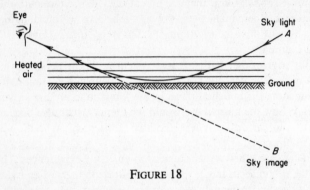

FIGURE 18

low down near the ground. As a result, the observer sees light from the sky originating at *A*, but it seems to come from *B*. This is precisely what would have happened if there had been a pool of water on the ground, for with wet ground we see the reflection of the sky's light. The effect of the heating of the road has therefore led to an apparent reflection of the kind that we always associate with water on the road. We are fooled and think the road is actually wet.

Most of the illusions we have examined were deliberately devised by psychologists, but we need not resort to contrived figures to appreciate that our sense of sight is constantly in error—and for understandable reasons. Because of the refraction or bending of light in the Earth's atmosphere, the sun is visible even when it is below the horizon. The Earth seems flat, and the sun seems to move around an apparently stationary Earth that does not even seem to rotate. Suppose the sun is high in the sky. To the question, "Do you see the sun now?" we would immediately answer yes. Yet the light from the sun takes eight minutes to reach us and in those eight minutes the sun could have exploded. When the sun is low on the horizon it does not appear as a circular disk but instead is somewhat flattened; its vertical diameter appears to be shortened. This phenomenon is caused by the bending of the light rays as they traverse the Earth's atmosphere. The stars, because they are so far away, seem like specks of light.

Visual distortions are often called illusions, but "illusions" come in many different forms. Color is transmitted to the brain from the retina through three channels. Three types of color receptors (cones) exist, each one sensitive to one of the three primary colors: red, green, or blue. White light activates all three color channels. Every object absorbs some light rays and reflects others. The color we see is what is reflected. A white object reflects all the light that falls on it. Is then a brown table actually brown? The flame from a candle in a brightly lit room seems dim, but in a dark room it seems bright. A piece of wood seems solid, but it is really a collection of atoms held together by interatomic forces. The hardness is not that of a continuous substance.

There may be distortions in other types of sensations: temperature, taste, the loudness or pitch of sound, and the speed at which objects appear to move. Let us consider an illusion of temperature. Immerse one hand in a bowl of hot water, the other in a bowl of cold water. After a few minutes, immerse both hands in a bowl of tepid water. Although both hands are now in water of the same temperature, the one that was in hot water feels the tepid as cold while the other feels it as hot. It is interesting to observe that if a hand is placed in water that is then heated (or cooled) gradually so that the change of temperature is not felt, the hand still adapts to the changed temperature.

The taste sense is also subject to several illusions. Sweet drinks taste gradually less sweet. Try keeping a strong solution of water and sugar in the mouth for a few seconds and then taste fresh water. It will now taste distinctly salty.

Misapprehensions of speed are common. A car moving at thirty miles per hour seems almost ridiculously slow after half an hour's continuous driving at highway speed. A very common illusion is presented by two trains in a station. If your train is stationary and the other is moving, you can readily be deceived into thinking that your train is also moving.

Some distortions are caused by the sensory receptors becoming fatigued or adapted by prolonged or intense stimulation. This can happen with any one of the senses and can result in considerable distortions. An example is the illusions of weight. After carrying a heavy weight for a few minutes, any much lighter weight will seem to be far lighter in proportion to what it actually is.

Beyond the illusions wherein we actually sense physical objects or happenings, we must take into account that our senses are limited. The normal human ear can hear sounds whose frequencies are about 20 to 20,000 cycles per second. The normal human eye can receive light whose wavelengths (see Chapter VII) range from 16 to 32 millionths of an inch. Yet both sound and light (strictly in the latter case, electromagnetic waves) exist and are physically real much beyond the ranges we can perceive with our senses alone. Even white light is not white but, as Newton showed, a composite of many frequencies. The eye registers only the composite. In fact, there are no colors in the physical world. As Goethe put it, color is what we see.

We never directly perceive a physical object but a sense datum. Our senses present and will always present not the faithful image of an objective reality, whether or not beyond our reach, but rather the image of the relationship between man and reality.

Humans claim, however, that beyond the senses we have intuition that can surely be relied on. Let us see how reliable human intuition is.

Suppose a man drives from New York City to Buffalo (a distance of 400 miles) at 60 miles per hour, and then drives back at 30 miles per hour. What is his average speed? Intuition almost certainly tells us that the average speed is 45 miles per hour. The correct answer, obtained by dividing the total distance by the total time, is about 40 miles per hour.

Let us consider some more examples of the reputed power of intuition. Suppose we put P dollars in the bank at a compound interest of i percent and keep it there until the total amount doubles. Let us suppose this happens in n years. It would seem reasonable to assume, if one put in $2P$ dollars at the same interest rate, that the $2P$ would double in fewer than n years. Actually it will take the same number of years for P and $2P$ to double.

Suppose a person rows two miles upstream and then two miles down a river that has a current of three miles per hour. Let us assume that the person can row five miles per hour in still water. How long should the entire trip take? Intuition suggests that the current will help the person as much on the way downstream as it will hinder on the way upstream. Hence the person will row four miles at five miles per hour, and the total time will be four-fifths of an hour. Actually the intuition is wrong, and the total time is one and a quarter hours.

Suppose one adds a quart of vermouth to a quart of gin to make a somewhat delectable martini. One should expect to obtain two quarts of martini. The correct answer, which certainly will elude the intuition, is one and nine-tenths quarts. Likewise, five pints of water and seven pints of alcohol do not make twelve pints of the mixture. In both cases molecules combine.

Let us now consider the problem of time. We can speak of the next second after a given second. A second is merely a duration of time. Intuition suggests that there is a next instant to a given instant. By instant we mean *no* duration of time, as for example the instant when a clock strikes one. But let us consider the paradox first presented by Zeno of Elea (fifth century B.C.). An arrow in flight is at one position at any one instant and at another position at the next one. How did the arrow have time to get to the next position?

Let us consider next a related problem of time. A clock strikes six in five seconds. How long should it take to strike twelve? Seemingly ten seconds. However, there are five intervals between the six strokes and eleven intervals between the twelve strokes. Hence the correct answer is eleven, not ten.

Let us consider a few more examples of intuition's failure. Consider two rectangles with the same perimeter. Must they have the same area? It would seem so. However, a little arithmetic soon tells us that this need not be so. We should then ask, of all rectangles with the same perimeter, which has the largest area? After all, if fencing is to be used to enclose the rectangle and if the area is to be used for planting, then the rectangle with the largest area is most desirable. The answer is a square.

A related problem asks us to consider two boxes of the same volume. Must the total area of the six sides of each be the same? Let us suppose that each box has a volume of 100 cubic feet. One can have dimensions of 50 by 1 by 2 feet, and the other 5 by 5 by 4 feet. The surface areas in this case are 204 and 130 square feet, respectively. Clearly, the difference is striking.

Another example of where intuition fails involves a young man who has a choice between two jobs. Each offers a starting salary of $1800 per year, but the first one would lead to an annual raise at the rate of $200, whereas the second would lead to a semiannual raise at the rate of $50. Which job is preferable? One would think that the answer is obvious. A raise of $200 per year seems better than one that apparently would amount to only $100 per year. But let us do a little arithmetic and consider what each job offers *during* successive *six-month* periods. The first job will pay 900, 900, 1000, 1000, 1100, 1100, 1200, 1200. . . . The second job, which provides a semiannual increase of $50, will pay 900, 950, 1000, 1050, 1100, 1150, 1200, 1250. . . .

It is clear from a comparison of these two sets of salaries that the second job brings a better return during the second half of each year and does as well as the first job during the first half. The second job is the better one. With the arithmetic before us it is possible to see more readily why the second job is better. The semiannual increase of $50 means that the salary will be higher at the rate of $50 for six months or at the rate of $100 for the year, because the recipient will get $50 more for each of the six-month periods. Hence two such increases per year amount to an increase at the rate of $200 per year. Thus far the two jobs seem to be equally good. However, on the second job the increases start after the first six months, whereas on the first job they do not start until after one year has elapsed. Hence the second job will pay more during the latter six months of each year.

Let us consider another simple problem. Suppose that a merchant sells his apples at two for five cents and his oranges at three for five cents. Being somewhat annoyed with having to do considerable arithmetic on each sale, the merchant decides to commingle apples and oranges and to sell any five pieces of fruit for ten cents. This move seems reasonable, because if he sells two apples and three oranges he sells five pieces of fruit and receives ten cents. Now he can charge two cents apiece, and his arithmetic on each sale is simple.

The dealer is cheating himself. Just to check quickly, we shall assume that he has one dozen apples and one dozen oranges for sale. If he sells apples normally at two for five cents, he receives thirty cents for the dozen apples. If he sells oranges at three for five cents, he receives twenty cents for the dozen oranges. His total receipts are then fifty cents. However, if he sells the twenty-four pieces at five for ten cents he will receive two cents per article or forty-eight cents.

The loss is due to poor reasoning on the part of the dealer. He assumed that the average price of the apples and oranges should be two cents each; however the average price per apple is two and one-half cents and the average price per orange one and two-thirds cents. The average price of two such items is two and one-twelfth cents per article and not two cents.

Next let us consider another common faulty intuition. Suppose we have a circular garden with a radius of 10 feet. We wish to protect the garden by a fence that is to be at each point 1 foot beyond the boundary of the garden. How much longer is the fence than the circumference of the garden itself? The answer is readily obtained. The circumference of the garden is given by a formula of geometry; this says that the circumference is 2π times the radius, π being the symbol for a number that is approximately 22/7. Hence the circumference of the garden is $2\pi \times 10$. The condition that the fence be 1 foot beyond the garden means that the radius of the circular fence is to be 11 feet. Hence the length of the fence is $2\pi \times 11$. The difference in these two circumferences is $22\pi - 20\pi$ or 2π. Therefore, the fence should be 2π feet longer than the circumference of the garden. There is nothing remarkable thus far.

We now consider a related problem. Suppose we were to build a roadway around the Earth—a trivial task for modern engineers—and the height of the roadway were to be 1 foot above the surface of the Earth all the way around. How much longer than the circumference of the Earth would the roadway be? Before calculating this quantity let us use our intuition at least to estimate it. The radius of the Earth is about 4000 miles or 21,120,000 feet. Since this radius is roughly two million times that of the garden we considered, one might expect that the *additional* length of the roadway should be about two million times the additional length of fence required to enclose the garden. The latter quantity was just 2π feet. Hence an intuitive argument for the additional length of roadway would seemingly lead to the figure of $2,000,000 \times 2\pi$ feet. Whether or not you would agree to this argument, you would almost certainly estimate that the length of the roadway would be very much greater than the circumference of the Earth.

A little mathematics tells the story. To avoid calculation with large numbers, let us denote the radius of the Earth by r. The circumference of the Earth is then $2\pi r$. The circumference or length of the roadway is $2\pi (r + 1)$. But the latter equals $2\pi r + 2\pi$. Hence the difference between the length of the roadway and the circumference of the Earth is just 2π feet, precisely the same figure that we obtained for the difference between the length of fence and the circumference of the garden, even though the roadway encircles an enormous Earth, whereas the fence encircles a small garden. In fact, the mathematics tells us even more. Regardless of what the value of r is, the difference, $2\pi (r + 1) - 2\pi r$, is always 2π, and this means that the circumference of the outer circle, if it is at each point 1 foot away from the inner circle, will always be just 2π feet longer than the circumference of the inner circle.

Intuition fails us in many other situations. A man some distance away from a tree notes that an apple is about to fall and wishes to hit the apple with a rifle bullet. He knows that by the time the bullet reaches the apple, it will have fallen some distance. Should he then aim at a point somewhat below the apple so that his bullet will hit it? No.

He should aim and fire at the apple. Both will fall the same distance downward during the time that the bullet is in flight.

As a final example of where intuition is likely to fail let us suppose there are 136 entrants in a tennis tournament and the director wishes to schedule a minimum number of matches to select the winner. How many need he schedule? Intuition seems helpless. The answer is 135, because each contestant must be defeated once, and once defeated, is eliminated.

Why are we subject to illusions of the senses and to false intuitions? Examination of the physiology of the various sensory organs could explain the sensory illusions, but for our purposes all we need is to recognize that the human sensory organs and the human brain are involved. With regard to intuition, it is actually a combination of experience, sense impressions, and crude guessing; at best one could say intuition is distilled experience. Subsequent analyses or experiments confirm or discredit it. Intuitions have been characterized as no more than force of habit rooted in psychological inertia.

When we speak of what is certain perceptually, we presuppose a separation between the perception and the perceiver. But this is impossible, because there can be no perception without a perceiver. What, then, is objective? We perhaps naively assume that what all perceivers agree on is objective. There are a sun and a moon. The sun is yellow and the moon is blue.

Helmholtz said in the *Handbook of Physiological Optics* (1896):

> It is easy to see that all the properties we ascribe to them [objects of the external world] signify only the effects they produce either on our senses or other external objects. Color, sound, taste, smell, temperature, smoothness, solidity belong to the first class; they signify effects on our sense organs. The chemical properties are likewise related to reactions, i.e., effects which the natural body in question exerts on others. It is thus with the other physical properties of bodies, the optical, the electric, the magnetic. . . . From this it follows that in fact, the properties of objects in nature do not signify, in spite of their name, anything proper to the objects in and for themselves but always a relation to a second body (including our sense organs).

What is our recourse to countering illusions and erroneous intuitions? The most effective answer is the use of mathematics. Just how effective the subject is remains to be seen. Our chief concern will be to show that there are phenomena in our physical world that are as real as any we perceive through our senses but that are extrasensory or not at all perceptible and, in fact, that in our present culture we utilize and rely on these extrasensory real phenomena, at least as much and per-haps even more than we rely on our sense perceptions.

This is not to say that mathematics does not utilize perceptions and intuitions as suggestions for its own development. However, mathematics surpasses these suggestions much as a diamond surpasses a piece of glass, and what mathematics reveals about our physical world is far more astonishing than the spectacle of the heavens.

II

The Rise and Role of Mathematics

In every specific natural science there can be found only so much science proper as there is mathematics present in it. Kant

The gods have not revealed all things from the beginning, but men seek and so find out better in time. Xenophanes

For the apparel oft proclaims the man. Shakespeare

Although the information obtained through our senses has been carefully observed, measured, and checked by experimentation, and although we now can utilize such aids as the telescope and the microscope, surveying instruments, and remarkably accurate measuring devices, the knowledge thus acquired is still limited and only approximately accurate. We know more about the number of planets, the presence of satellites of several planets, dark spots on the sun, and the use of the compass to navigate the seas. Yet all of these gains in knowledge are insignificant in comparison with the variety and importance of the phenomena we need to and wish to study.

The crucial, powerful, and decisive step that has increased and enhanced our knowledge of the physical world is the employment of mathematics. The role of this tool is so far superior to the means just described that it can be labeled superlative and even miraculous. Not only does it correct and increase our knowledge of phenomena that are perceptible, it also reveals vital phenomena that are not at all perceptible but that are as real in their effects as touching a hot stove. There are physical "ghosts" whose presence in our daily lives cannot be doubted. How their existence was disclosed will be our next focus of discussion.

To those of us who were educated in Western Europe and the Americas, the nature and mundane uses of mathematics are familiar and accepted as almost commonplace. Even the civilizations we credit as originating Western European mathematics, namely, the Babylonian and Egyptian civilizations, produced from about 3000 B.C. a collection of useful but disconnected rules and formulas to solve such practical problems as humans encounter in their everyday life. These peoples did not recognize the power of mathematics to extend their knowledge of nature beyond what the senses reveal. Their mathematics may be regarded as the alchemy that preceded chemistry.

Mathematics, as a logical development and a tool to learn more about nature, is a creation of the Greeks that they began to take seriously about 600 B.C. There are no surviving documents of the sixth and fifth centuries B.C. that can tell us how or why the Greeks came to this new concept and role of mathematics. Instead, we have the speculations of historians, one of whom states that the Greeks found contradictory results on the area of a circle in the works of Babylonians and Egyptians and so had to decide which was correct. There were similar disagreements on other topics. Another explanation cites the philosophical interests of the Greeks, but this suggestion raises more questions than it answers. Still another sees deductive mathematics originating from Aristotelian logic, which arose from disputation on political and social issues. But Greek mathematics precedes Aristotle.

Perhaps all one can say is that from the sixth century B.C. onward the Greeks had a vision, the substance of which was that nature is rationally designed and that all natural phenomena follow a precise and unvarying plan, indeed a mathematical plan. The human mind has superb power, and if this power is applied to the study of nature, the rational, mathematical pattern can be discerned and rendered intelligible.

In any case the Greeks became the first people with the audacity and the genius to give reasoned explanations of natural phenomena. The Greek urge to understand had the excitement of a quest and an exploration. While they explored they made maps, such as Euclidean geometry, so that others might find their way quickly to the frontiers and help conquer new regions.

We are on somewhat safer historical grounds when we cite that Thales (c. 640–c. 546 B.C.), who lived in the Greek city of Miletus in Asia Minor, proved several theorems of Euclidean geometry. There are no documents of his time, and the belief that he proved theorems by logical means is somewhat dubious; yet it is certain that he and his contemporaries in Asia Minor speculated on the design of nature.

We are more certain that with the advent of the Pythagoreans, a mystical and religious order of the sixth century B.C., the program of determining the rational design of nature enlisted the aid of mathematics. The Pythagoreans were struck by the fact that phenomena that are physically very diverse exhibit identical mathematical properties. The moon and a rubber ball share the same shape and many other properties common to spheres. Was it not apparent, then, that mathematical relations underlie diversity and must be the essence of phenomena?

To be specific, the Pythagoreans found this essence in number and in numerical relations. Number was the first principle in the description of nature, and it was the matter and form of the universe. The Pythagoreans are reported to have believed that "all things are numbers." This belief makes more sense when we take into account that the Pythagoreans visualized numbers as dots (which could have meant particles to them) and that they arranged dots in patterns, each of which could be taken to represent a real object. Thus the collections

$$\begin{matrix} & & \cdot & & & \cdot & \cdot & \cdot \\ \cdot & & \cdot & \text{and} & & \cdot & \cdot & \cdot \\ & & & & & \cdot & \cdot & \cdot \end{matrix}$$

were called triangular and square numbers, and may have been regarded as representing triangular and square objects. There is no doubt that as the Pythagoreans developed and refined their own doctrines they began to understand numbers as abstract concepts and physical objects as concrete realizations.

The Pythagoreans are credited with the reduction of music to simple relationships among numbers when they discovered two facts: first, the sound caused by a plucked string depends on the length of the string; and second, harmonious sounds are given off by strings whose lengths can be expressed as ratios of whole numbers. For example, a harmonious sound is produced by plucking two equally taut strings, one twice as long as the other. The musical interval between the two notes is now called an octave. Another harmonious combination is formed by plucking two strings whose lengths are in the ratio of three to two; in this case the shorter one gives forth a note called the fifth above that given off by the longer one. In fact, the relative lengths in every harmonious combination of plucked strings can be expressed as ratios of whole numbers.

The Pythagoreans also "reduced" the motions of the planets to numerical relations. They believed that bodies moving in space produce sounds and that a body moving rapidly gives forth a higher note than one moving more slowly. Perhaps these ideas were suggested by the swishing sound of an object whirled on the end of a string. According to Pythagorean astronomy, the greater the distance of a planet from the Earth the more rapidly it moved. Hence the sounds produced by the planets varied with their distance from the Earth, and these sounds all harmonized. But this "music of the spheres," like all harmony, could be reduced to mere numerical relationships, and hence so could the motions of the planets.

Other features of nature were "reduced" to number. The numbers 1, 2, 3, 4, the *tetractys,* were especially valued. In fact, the Pythagorean oath is reported to be: "I swear in the name of the Tetractys which has been bestowed on our soul. The source and roots of the overflowing nature are contained in it." Nature was composed of "fournesses," such as the four geometrical elements (point, line, surface, and solid), and the four material elements Plato later emphasized (earth, air, fire, and water).

The four numbers of the *tetractys* added up to ten, and so ten was the ideal number and represented the universe. Because ten was ideal there must be ten bodies in the heavens. To specify the required number of bodies the Pythagoreans introduced a central fire around which the Earth, sun, moon, and the five planets then known revolved, and a counter-earth on the opposite side of the central fire. Neither this central fire nor the counter-earth is visible because the area of the Earth on which we live faces away from them. In this way the Pythagoreans built an astronomical theory based on numerical relationships.

With these examples we can make sense of the statement attributed to Philolaus, a famous fifth-century B.C. Pythagorean:

> Were it not for number and its nature nothing that exists would be clear to anybody either in itself or in its relation to other things. . . . You can observe the power of number exercising itself not only in the affairs of demons and gods but in all the acts and thoughts of men, in all handicrafts and music.

The natural philosophy of the Pythagoreans is hardly substantial. Moreover, the Pythagoreans did not develop any one branch of physical science very far. Justifiably, one could call their theories superfi-

cial. Whether by luck or by intuitive genius, however, the Pythagoreans did advance two doctrines that proved later to be all-important:
the first is that nature is built according to mathematical principles,
and the second, that number relationships underlie, unify, and reveal
the order in nature.

The atomists Leucippus (*c.* 440 B.C.) and Democritus (*c.* 460–
c. 370 B.C.) also advanced the importance of mathematics. They
believed that all matter consists of atoms that differ in position, size,
and shape. These were physically real properties of the atoms. All other
properties such as taste, heat, and color were not in the atoms but in
the effect of the atoms on the perceiver. This sensuous knowledge was
unreliable because it varied with the perceiver. Like the Pythagoreans,
the atomists asserted that the reality underlying the constantly changing features of the physical world was expressible in terms of mathematics. Thus, the happenings in this world were strictly determined by
mathematical laws.

The Greek who most effectively promoted the mathematical
investigation of nature was Plato (427–347 B.C.). Plato took over some
Pythagorean doctrines but was a master in his own right who dominated Greek thought in the momentous fourth century B.C. He was the
founder of the Academy in Athens, a center that attracted the leading
thinkers of his day and that endured for nine hundred years. His views
are clearly expressed in his dialogue *Philebus.* We have already noted
(see Historical Overview) that the real world according to Plato was
designed mathematically. What we perceive through our senses is an
imperfect representation of the real world. The reality and intelligibility of the physical world could be comprehended only through mathematics, for "God eternally geometrizes." Plato went further than
most Pythagoreans in that he wished not merely to understand nature
through mathematics but to transcend nature to comprehend the ideal,
mathematically organized world that he believed to be the true reality.
The sensory, the impermanent, and the imperfect were to be replaced
by the abstract, eternal, and perfect. He hoped that a few penetrating
observations of the physical world would suggest basic truths that
could then be developed through reason; at this point, he could dis-
pense with further observation. And from here on, nature would be
replaced entirely by mathematics. Indeed, he criticized the Pythagoreans because they investigated the numbers of the harmonies that are
heard but never reached the natural harmonies of numbers themselves. For Plato, mathematics was not only the mediator between the
ideas and the things of sense; the mathematical order was the true

account of the nature of reality. Plato also laid down the principles of the axiomatic-deductive method, which we shall discuss shortly. He saw this method as the ideal way of systematizing knowledge and arriving at new knowledge.

The pursuit of mathematics to study and obtain true knowledge of our physical world was urged also by the leading successor of Plato. Although Aristotle and his followers differed somewhat from the Platonists concerning the relationship of mathematics to the real world, this school also expounded and advocated the mathematical design of nature. Aristotle affirmed that the abstractions of mathematics were derived from the material world; however, there are no passages in his writings that advocate mathematics as a correction or extension of sensory knowledge. He did believe that the motions of the heavenly bodies were mathematically designed but, basically, that mathematical laws were merely a description of events. For Aristotle the final cause or objective of events, the teleological doctrine, was most important.

When Alexander the Great (356–323 B.C.) set out to conquer the world, he transferred the center of the Greek world from Athens to a city of Egypt, which he modestly named Alexandria. It was in Alexandria that Euclid (c. 300 B.C.) wrote the first memorable document on mathematical knowledge, the classic *Elements.* Here proof makes its first known appearance. Euclid also wrote tracts on mechanics, optics, and music in which mathematics was the core; mathematics was the ideal version of what the known physical world contained. Some of his theorems indeed offered new knowledge of geometrical figures and of properties of the whole numbers. However, because we have no original manuscripts by Euclid we do not know whether new knowledge was his objective or whether he was concerned with the reliability of sensory knowledge. In any case he led the way for other creators of mathematics.

The Greeks of the Alexandrian period (c. 300 B.C.–A.D. 600) extended mathematics almost immeasurably. For present purposes we need only note the major work by Apollonius (c. 262–c. 190 B.C.), *Conic Sections;* a variety of first-class works by Archimedes (c. 287–212 B.C.) on many areas of mathematics and mechanics; the work on trigonometry by Hipparchus, Menelaus, and Ptolemy (c. A.D. 85–c. 165); and, late in the period (c. A.D. 250), the arithmetical work of Diophantus. All of these works, like Euclid's, gave ideal versions of objects, relationships, and phenomena of the physical world and extended our knowledge.

The Greek civilization was destroyed by the conquests of the Romans and Mohammedans, and with its demise Europe entered the Middle Ages, which endured for a thousand years from about A.D. 500 to 1500. This culture was dominated by the Catholic church, which subordinated life on Earth to preparation for an afterlife in heaven. Consequently, the study of nature by any means, mathematical or otherwise, was deprecated. Nevertheless, a few individuals and groups (Robert Grosseteste, Roger Bacon, John Peckham, the Mertonians at Oxford—among whose members were William of Ockham, Thomas Bradwardine, Abelard of Bath, Thierry of Chartres, and William of Conches) did make some efforts to continue mathematical and physical investigation. In particular they subscribed to mathematics as the veridical account of physical phenomena, and some, notably Abelard and Thierry, insisted also on experimental techniques. All of these thinkers believed that the universe was basically rational and that mathematical reasoning could produce knowledge about it. Nor should we overlook during this medieval period the contributions of the Hindus and the Arabs, which were gradually absorbed into the body of mathematics.

The modern period, our main concern, may be thought to begin about 1500. The sixteenth century in particular is often distinguished as the Renaissance, the rebirth of Greek thought. Just how Greek manuscripts reached Italy, the center of the Renaissance, is irrelevant to our account. It may suffice to say that the Greek ideas fascinated the Europeans.

The Europeans generally did not respond immediately to the new forces and influences. During the period often labeled "humanistic," the study of Greek works was far more characteristic than the active pursuit of Greek objectives, but by about A.D. 1500 European minds, infused with Greek goals—that is, the application of reason to the study of nature and the search for the underlying mathematical design—began to act. However, they faced a serious problem in that the Greek goals were in conflict with the prevailing culture. Whereas the Greeks believed in the mathematical design of nature, with nature conforming invariably and unalterably to some ideal plan, late medieval thinkers ascribed all plan and action to the Christian God. He was the designer and creator, and all the actions of nature followed the plan laid down by this agency. The universe was the handiwork of God and subject to His will. The mathematicians and scientists of the Renaissance and several succeeding centuries were orthodox Christians and so accepted this doctrine. But Catholic teachings by no means included the Greek doctrine of the mathematical design of nature. How, then, was the attempt to understand God's universe to be reconciled with the search for the mathematical laws of nature? The answer was to add a new doctrine—that the Christian God had designed the universe mathematically. Thus, the Catholic doctrine postulating the supreme importance of seeking to understand God's will and His creations took the form of a search for God's mathematical design of nature. Indeed, the work of sixteenth-, seventeenth-, and most eighteenth-century mathematicians was, as we shall soon see more clearly, a religious quest. The search for the mathematical laws of nature was an act of devotion that would reveal the glory and grandeur of His handiwork.

Mathematical knowledge, the truth about God's design of the universe, thus became as sacrosanct as any line of Scripture. Humans could not hope to perceive the divine plan as clearly as God Himself understood it, but humans could with humility and modesty seek at least to approach the mind of God and so understand God's world.

One can go further and assert that these mathematicians were sure of the existence of mathematical laws underlying natural phenomena and persisted in the search for them because they were convinced a priori that God had incorporated them into the construction of the universe. Each discovery of a law of nature was hailed as evidence of God's brilliance rather than that of the investigator. The beliefs and attitudes of the mathematicians and scientists swept Renaissance Europe. The recently discovered Greek works confronted a deeply devout Christian world, and the intellectual leaders born in one word and attracted to the other fused the doctrines of both.

Alongside this new intellectual fervor, another doctrine was gaining support—the idea of "back to nature." Every variety of scientist abandoned endless rationalizing on the basis of dogmatic principles, vague in meaning and unrelated to experience, and turned to nature herself as the true source of knowledge. Certainly by 1600 the Europeans were motivated to undertake what has often been described as the Scientific Revolution. Several happenings motivated or accelerated this movement: geographical explorations disclosed new lands and new peoples; the invention of the telescope and microscope revealed new phenomena; the compass aided navigation; the heliocentric theory introduced by Copernicus (see Chapter IV) stimulated new thoughts about our planetary system; and the Protestant Revolution challenged Catholic doctrines. Mathematics soon resumed its major role as the key to nature.

This brief sketch of the historical background of modern European mathematics is intended primarily to indicate that mathematics and its uses in the investigation of nature, which will be our main concern in succeeding chapters, did not originate as a bolt from the blue. However, our concern will not be the elementary mathematics that provided the tools to correct and extend our knowledge of commonly perceptible phenomena but, instead, what mathematics has achieved in revealing and describing phenomena that either are not readily accessible or are not at all so through the senses. For our purposes we need not pursue and master the techniques of mathematics, but it is essential to understand how mathematics enables us to represent physical phenomena and to arrive at knowledge about these phenomena.

What are the essential elements of the mathematical method? The first is the introduction of basic concepts. Some, such as point, line, and whole number, are suggested directly by material or physical objects. Beyond the elementary concepts, mathematics has in fact become dominated by concepts derived from the recesses of the human mind. To cite a few examples of such concepts: negative numbers, letters standing for classes of numbers, complex numbers, functions, all sorts of curves, infinite series, concepts of the calculus, differential equations, matrices and groups, and higher dimensional spaces.

Some of the above concepts lack entirely an intuitive meaning. Others do have some intuitive basis in physical phenomena as, for example, the *derivative* or instantaneous rate of change. However, although it is related to the physical phenomenon of velocity, the derivative is far more an intellectual construct and is qualitatively an entirely different sort of contribution from that of the mathematical triangle.

Throughout the history of mathematics new concepts were viewed with suspicion at the outset. Even the notion of negative numbers was originally rejected by serious mathematicians. However, each new concept was grudgingly accepted as its usefulness in application became evident.

A second essential feature of mathematics is abstraction. Speaking of geometricians, Plato said in the *Republic:*

> Do you not know also that although they make use of the visible forms and reason about them, they are thinking not of these, but of the ideals which they resemble; not of the figures which they draw, but of absolute square and the absolute diameter . . . they are really seeking to behold the things themselves which can be seen only with the eye of the mind?

If mathematics is to be powerful, it must embrace in one abstract concept the essential features of all the physical manifestations of that concept. Thus, the mathematical straight line must embrace stretched strings, rulers' edges, boundaries of fields, and the paths of light rays.

That the concepts are abstractions is exemplified in the most elementary concept, *number*. Failure to recognize this can lead to confusion. A simple situation can be used to make the point. A man goes into a shoestore and buys three pairs of shoes at $20 a pair. The salesperson says that three pairs of shoes at $20 a pair cost $60 and he expects the customer to hand him $60. But the customer instead replies that three pairs of shoes at $20 a pair is not $60 but sixty pairs of shoes, and he asks the salesperson for the sixty pairs. Is the customer right? As right as the salesperson. If pairs of shoes times dollars can yield dollars, then why cannot the same product yield pairs of shoes? The answer is, of course, that we do not multiply shoes by dollars. We abstract the numbers three and twenty from the physical situation, multiply to obtain sixty and then interpret the result to suit the physical situation.

Another feature of mathematics is idealization. A mathematician idealizes by deliberately ignoring the thickness of a chalk mark in treating straight lines or by regarding the Earth as a perfect sphere in some problems. Idealization per se is not a serious departure from reality, but it does raise the question in any application to reality whether the real particle or path under study is close enough to its idealization.

The most striking feature of mathematics is the method of reasoning it employs. The basis is a set of axioms and the application of deductive reasoning to these axioms. The word *axiom* comes from the Greek, meaning "to think worthy." The Greeks introduced the notion of axioms—truths so self-evident that no one could doubt them. Plato's theory of *anamnesis* stated that humans had a prior experience of truth as souls in an objective world of truths and that the axioms of geometry represented the recollection of previously known truths. Aristotle maintained in *Posterior Analytics* that the axioms are known to be true by our infallible intuition. Moreover, we must have these truths on which to base our reasoning. If, instead, reasoning were to use some facts not known to be truths, further reasoning would be needed to establish these facts, and this process would have to be repeated endlessly. Aristotle also pointed out that some concepts must remain undefined or else there would be no starting point. Today such terms as point and straight line are undefined; their meaning and properties depend on the axioms that prescribe their properties.

Just as many of the concepts with which mathematics deals are invented by human minds, so the axioms about these concepts are invented to suit what the concepts are intended to reveal about reality. Thus, axioms for negative and complex numbers must necessarily be different from those for positive numbers, or at least the latter must be extended to include negative and complex numbers. The subtleties in the newer concepts are far greater, and the correct axiomatic bases for some branches of mathematics were achieved long after the branch was established.

Beyond mathematical axioms, some physical knowledge must enter into major contributions of mathematics to our physical world. This may take the form of physical axioms such as Newton's laws of motion, generalizations of experimental observations, or sheer intuition. These physical assumptions are formulated in the language of mathematics and so permit the axioms and theorems of mathematics to be applied to them.

However basic the concepts and axioms, it is the *deductions* from the axioms that allow us to acquire totally new knowledge to correct our sense perceptions. Of the many types of reasoning—for example, inductive, analogical, and deductive—only deductive guarantees the correctness of the conclusion. To conclude that all apples are red because 1000 apples are found to be red is inductive reasoning, therefore not reliable. Similarly, the argument that John should be able to graduate from college because his identical twin who inherited the same faculties did so, is reasoning by analogy, and is certainly not reliable. Deductive reasoning, on the other hand, although it can take many forms, does guarantee the conclusion. Thus, if one grants that all men are mortal and Socrates is a man, one must accept that Socrates is mortal. The principle of logic involved here is one form of what Aristotle called syllogistic reasoning. Among other laws of deductive reasoning Aristotle included the law of contradiction (a proposition cannot be both true and false) and the law of excluded middle (a proposition must be either true or false).

He and the world at large accepted unquestioningly that these deductive principles when applied to any premises yielded conclusions as reliable as the premises. Hence, if the premises were truths, so would be the conclusions. It is worthy of note that Aristotle abstracted the principles of deductive logic from the reasoning already practiced by mathematicians. Deductive logic is, in effect, the child of mathematics.

It is important to appreciate how radical the insistence on deductive proof is. We can test as many even numbers as we wish and find that each is a sum of two prime numbers. However, we can not state that this result is a mathematical theorem because it was not obtained by a deductive proof. Similarly, suppose a scientist were to measure the sum of the angles of 100 different triangles in different locations and of different size and shape, and find that sum to be 180 degrees to within the limits of experimental accuracy. Surely this scientist would conclude that the sum of the angles of any triangle is 180 degrees. However, not only were the measurements approximate, there remained the question of whether some triangular shape not measured would produce a markedly different result. The scientist's inductive proof is not mathematically acceptable. The mathematician, on the other hand, starts with facts or axioms that seem to be reliable. Who can doubt that if equals be added to equals, the sums are equal? By means of such indubitable axioms one can prove deductively that the sum of the angles of *any* triangle is 180 degrees.

The deductive process we have described uses logic to justify the reasoning. What has been employed practically up to modern times is what is called Aristotelian logic. We may ask why the conclusions derived by this application of logic should apply to nature. Why should theorems deduced by human minds sitting in cloistered rooms be as applicable to the real world as the axioms that are themselves in many cases also suggested only by human minds? We shall return to the question of why mathematics works in Chapter XII.

We have yet to cite another important feature of mathematics— the use of symbolism. Although a page of mathematical symbols can hardly be described as appealing, there is no question that without symbolism mathematicians would be lost in a wilderness of words. All of us use symbolism in a host of common abbreviations. We use N.Y. to mean New York, for example, and although the meaning of such symbols must be learned, there is no question that the brevity of symbolism permits comprehension, whereas a verbal expression would overburden the mind.

We can sum up the means by which mathematics derives facts about our physical world by saying that it builds models for classes of real phenomena. Concepts, usually idealized (whether drawn from observation of nature or supplied by human minds); axioms, which may also be suggested by physical facts or by humans; and the processes of idealization, generalization, and abstraction, as well as intuition, are all utilized in the building of models. Proof, of course, cements the components of a model. The most familiar model is Euclidean geometry, but we shall examine many more sophisticated and more ingenious models that tell us far more about far less obvious phenomena than Euclidean geometry.

Our goal, then, is to see how firmly mathematics enters the modern world, not just as the method of correcting the imperfections of the senses but more especially as a method of extending the knowledge humans can acquire about their world. As Hamlet put it, "There are more things in Heaven and earth, Horatio, than are dreamt of in your philosophy." We must go beyond perceptual knowledge. The essence of mathematics, as opposed to sense perceptions, is that it draws on the human mind and human reasoning to produce knowledge about our physical world that the average human being, even in Western culture, believes is acquired entirely by the use of sense perceptions.

Alfred North Whitehead, in his *Science and the Modern World,* has emphasized the importance of mathematics in the exploration of our physical world.

> Nothing is more impressive than the fact that as mathematics withdrew increasingly into the upper regions of ever greater abstract thought, it returned back to earth with a corresponding growth of importance for the analysis of concrete fact. . . . The paradox is now fully established that the utmost abstractions are the true weapon with which to control our thought of concrete fact.

And as David Hilbert, the foremost twentieth-century mathematician, remarked, physics is now too important to be left to the physicists.

III

The Astronomical Worlds of the Greeks

> SOCRATES: *Very good; let us begin then, Protarchus, by asking a question.*
>
> PROTARCHUS: *What question?*
>
> SOCRATES: *Whether all this which they call the universe is left to the guidance of unreason and chance medley, or, on the contrary, as our fathers have declared, ordered and governed by a marvellous intelligence and wisdom.*
>
> PROTARCHUS: *Wide asunder are the two assertions, illustrious Socrates, for that which you were just now saying to me appears to be blasphemy, but the other assertion, that mind orders all things, is worthy of the aspect of the world....*
>
> Plato, *Philebus*

As we all know, the astronomical theories of the Greeks did not survive. Yet they are the earliest prime examples of how mathematics made sense of the world of perceptions. Moreover, we can appreciate all the more the magnitude of the revolution in astronomy initiated by Copernicus and Kepler when we view it in the light of what preceded.

We are concerned here with what mathematics reveals about our physical world that is not perceptible or at best perceptible but so inadequately that our perceptions are gross misrepresentations of what is physically real and significant. In such applications the Greeks excelled in mathematical astronomy and paved the way for far more successful mathematical theories.

The basic reason for their emphasis on the astronomical work is that the heavens presented the most complicated motions, at least insofar as the human eye could discern. There were no telescopes in

Grecian times, and even if there had been it is unlikely that that instrument would have been sufficiently helpful to determine some patterns in the heavenly motions. The appearances, disappearances, and reappearances of stars and starlike bodies were distressing and mysterious.

Although the Greeks did not provide the current mathematical astronomy, they initiated it and indeed provided suggestions for the theory that superseded it. The first truly mathematical reasoning and understanding of cosmic phenomena were initiated by the Greeks.

Interest in the heavenly bodies did exist even in the most primitive societies. The light and heat of the sun, the spectacular colors that the sun and moon often assume, the bright lights of the planets that appear and disappear at various times of the year, the amazing panorama of lights from the Milky Way, and eclipses excited wonder, admiration, speculation, and, in some cases, terror. However, any somewhat accurate knowledge about these phenomena was limited in pre-Grecian times to the periods of revolution of the sun and moon, and to the times of appearance and disappearance of some planets and stars. This information was unfortunately inadequate to yield any estimates of the sizes and distances of these bodies, and still less adequate to furnish any account of their relative motions.

The Egyptians and the Babylonians did make observations primarily of the motions of the sun and moon, partly for calendar reckoning and partly for knowledge of the seasons, which was important for agriculture. However, neither of these peoples nor other cultures preceding the Greeks ever constructed a comprehensive account of the motions of the heavenly bodies. Certainly they lacked the requisite mathematical knowledge and had no really effective observational instruments. The complex behavior of the heavenly bodies concealed from them any indication of plan, order, and law. Nature appeared capricious and mysterious.

The Greeks thought otherwise. Spurred on by their desire for knowledge and their love of reason, they were confident that an examination of nature's ways would reveal the order inherent in the heavenly motions. We shall see that many of the Greek astronomers advanced and defended ideas that ultimately became part of modern cosmology. This cosmology did not result from a single genius. If genius was involved it was the work of a succession of geniuses.

The study of the heavens began in Miletus, southernmost of the twelve cities of Ionia on the western border of Asia Minor. Here, in the sixth century before our era, a classic combination of the factors occurred that liberated the human intellect to partake in the joyful and often dangerous pursuit of speculation. Industry and trade had brought wealth to the city and had blessed its inhabitants with comfort and leisure. The citizens traveled widely, imbibing from Egypt and Babylonia and elsewhere the full richness of Oriental thought. The Milesians saw their material prosperity as evidence of what they could accomplish unaided by the gods; and gradually a few bold spirits dared to believe that the universe itself is an intelligible whole, accessible to the human mind.

To Thales belongs the dual honor of standing first among both the scientists and the philosophers of the Western tradition. His devotion to the heavens is supported by the story of his falling into a ditch while studying the stars. He won renown by allegedly forecasting the solar eclipse of 585 B.C., but modern historians cast doubt on the prediction.

The successors of Thales, Anaximander (611–549 B.C.) and Anaximenes (570–480 B.C.), continued to speculate and advance theories about the fundamental matter of the universe and its structure. Mathematics played no essential role in these speculations, however. In the absence of instruments and any established methodology, these scientists could only guess about the nature of the heavenly bodies and about their distances from the Earth; thus even Anaximander supposed that the stars are nearer to us than the sun and moon. No mention was made of the planets as such; these "wanderers" (the word planet means wanderer in Greek) were still regarded as essentially similar to the other stars.

Nevertheless, Thales and his Ionian colleagues progressed far beyond the thinking of preceding civilizations. At the very least, these men dared to tackle the universe, and they refused any help from gods, spirits, ghosts, devils, angels, or other agents unacceptable to a rational mind. Their material and objective explanations and their reasoned approaches discredited the fanciful and uncritical explanations of the poetic, mythical, and supernatural accounts. Brilliant intuitions fathomed the nature of the universe and reason defended these insights.

In the philosophy and science of the Greeks, the next major figure was Pythagoras (sixth century B.C.). Pythagoras was born on the island of Samos, near Miletus. He broadened himself through thirty years of travel, fled Samos to escape a political tyranny, and came at last, at about the age of fifty, to Crotona, Italy. There he gathered his disciples into a strange brotherhood whose initiates combined scientific study with religious ritual.

In the realm of astronomy, the Pythagorean doctrine revolutionized cosmology by boldly declaring that the Earth is a sphere. The master himself so believed, and the new idea was committed to writing by Parmenides (*c.* 500 B.C.). Apparently the motives of these thinkers were as much aesthetic as scientific. Pythagoras regarded the sphere as the most beautiful of solid objects; he taught that the universe itself has this perfect form, and he seems to have felt that heaven and Earth should share a common shape. Such feelings may have been prompted or at least bolstered by reports from observant seafarers and by the observations made at the time of eclipses. Gradually the sphericity of the Earth won general acceptance, although Aristotle, writing in the middle of the fourth century, implies that the disagreement had not then subsided.

The Pythagoreans did create a cosmology, but it was purely speculative and had little influence on later Greek astronomical thought. Its number mysticism and its a priori character may seem unscientific—until we remember the rudimentary state of observational astronomy. We shall see that in general the Greek astronomers sensed vividly the inevitable imprecision of their sky-watching and turned to mathematics as a far more reliable road to celestial certainty.

It was the perversely irregular motions of the planets that now commanded the attention of the astronomers. Gradually, to be sure, a few pieces of the puzzle began falling into place. The stargazers realized that Venus and Mercury, unlike the other three "wanderers" then known, remain always close to the sun and therefore can be seen only in the morning or evening; and they learned to identify the "morning star" and the "evening star" as one and the same. Meanwhile, they observed and pondered the mystery of planetary retrogradation—the strange way that the wanderers would sometimes halt in their normal west-to-east course across the sky, remain stationary for a time, retrace their paths in an east-to-west direction for a short distance, pause again, and finally resume their eastward motion. This erratic behavior became the despair of astronomers, and the sensitive Greek spirit, which loved order and regularity, was almost horrified by these celestial vagabonds. Might there not be, after all, some pattern underlying this apparent chaos?

Now it is one thing to observe and carefully chart the motions of the planets as the Egyptians and Babylonians did for centuries. These people were merely observers. It is quite another, and indeed a major step forward, to ask for some unifying theory of the motions of the heavenly bodies that will reveal a plan underlying the seeming irregularity. This is the problem Plato set before the Academy in the now famous phrase "saving the appearances." An answer to Plato's problem given by Eudoxus, a pupil of Plato, a master in his own right and one of the foremost Greek mathematicians, is the first major astronomical theory known to history and a decided advance in the program of rationalizing nature.

Eudoxus (c. 408–355 B.C.) came from Cnidus, on the west coast of Turkey. As a young man he traveled to Italy and Sicily to study geometry under Archytas, and there he rose to eminence for his mathematical theory. At twenty-two he journeyed to Athens where he attended Plato's lectures at the Academy. He made observations of his own, and centuries later his "observatory" was still pointed out to curious travelers.

Eudoxus' scheme employed a series of concentric spheres whose center is the immovable Earth. To account for the complex motion of any one body other than the immovable Earth, Eudoxus supposed that a combination of spherical motions would produce the desired path. The scheme is rather detailed, because the system of three or four spheres needed for any one body varies from planet to planet and for the sun and the moon. These spheres were of course only mathematical and hypothetical.

With these achievements Eudoxus seems to have been content. He made no inquiry into the physical nature of his spheres or of their interconnections, or the physical cause of their motions. It is reasonable to assume that he saw his system as no more than a beautiful theory, neither requiring nor presuming physical verification. Such a stance, if really taken by Eudoxus, would place him at the head of a durable tradition in ancient, medieval, and modern astronomy that regarded geometrical models of the heavens only as convenient mathematical fictions.

How well did Eudoxus' theory actually represent the motions observed in the heavens? His own writings are lost, and his scheme is known to us only through the accounts of ancient commentators, notably Aristotle. The usual assessment has concluded that appropriate combinations of spheres will yield quite accurately all relevant motions except the paths of Venus and Mars. For these two bodies, by contrast, the system breaks down badly, failing even to provide the retrogradations that are the single most striking feature of planetary motion. In antiquity, however, it was another defect that formed the greater basis of opposition to the scheme. Critics argued that if, as Eudoxus supposed, the heavenly bodies keep always the same distance from Earth, they should not show the variations in brightness or size that can be conspicuous even to the naked eye. Eudoxus himself had been aware of the difficulties and saw fit to ignore them. Doubtless he and his contemporaries saw clearly enough that the overthrow of the theory might dislodge Earth from the center of the cosmos; and from so awesome an outcome, all but the bravest spirits shrank.

Even as the Pythagoreans, the Stoics, the Epicureans, Plato, and Aristotle were still maintaining that theories of the universe were not yet the private preserve of science, a few gifted people were pursuing ideas that would ultimately turn cosmology into a study accessible only through the arcane language of mathematics. In the middle of the fourth century B.C. a certain Heraclides, called "Ponticus" for his place of birth, put forth two suggestions of revolutionary import. The apparent daily rotation of the heavens, said Heraclides, is an illusion. In reality it is the Earth that moves, spinning on its axis once in twenty-four hours. This was a daring speculation, for like most great scientific advances it flew in the face of common sense and sensory experience. Actually, the rotation of the Earth had already been proclaimed two centuries earlier by the obscure Pythagorean philosopher Hiketas, and perhaps the idea had never quite died. People must have always realized that either of the two possibilities—the revolution of the sky or the rotation of the Earth—would fit the observations equally well. Why, then, did Heraclides choose to set the Earth in motion? Perhaps Heraclides, sharing the general awareness that the cosmos must be very large compared to the Earth, preferred spinning our tiny globe to revolving the vastness of the periphery. Whatever its rationale, the new idea did not find immediate or general acceptance.

His other innovation was even more far-reaching, for it attacked the prevailing cosmology at its most vulnerable point. We have seen that Eudoxus' system of concentric spheres could not explain the observed variation in the size or brightness of celestial objects. Despite this drawback, the theory continued for a time to find adherents eager for its defense.

Not until Heraclides pointed the way was it possible to see alternatives. Presumably motivated by the constant nearness of Venus and Mercury to the sun, he suggested that these two planets travel in circles with the sun as center. If, said Heraclides, this "heliocentric" motion occurs in combination with the sun's own circular course around Earth, the distances from us of Venus and Mercury will obviously vary, producing just the sort of fluctuating brightness that the theory of Eudoxus was powerless to explain. The new hypothesis had an immense and lasting impact. On the purely mathematical side, Heraclides had conjured up, for the first time in cosmological speculation, the idea of an "epicycle": a circle whose center moves on another circle. Much history would stem from this beginning. His theory was weakened by its limitation to two of the five planets, a restriction that Heraclides apparently did not seek to remove; but the notion of the sun as a center of celestial movement was another milestone in the long erosion of the favored position that naive observation and human pride bestowed on our Earth.

Interest in quantitative knowledge and the willingness to accumulate this knowledge finally arose in the second great Greek period, when the center of that civilization was moved to Alexandria. It is irrelevant to trace here the changes in the character of the Greek civilization that took place in the Alexandrian period. What is important, however, is that in that city the Greeks came into close contact with Egyptians and Babylonians, and the wealth of astronomical observations acquired by the Egyptians and Babylonians over several millenniums became more accessible. Equally pertinent is the fact that those successors of Alexander who ruled the Egyptian empire built a great home for scholars, called the Museum, and spent lavishly to equip a famous library. They also provided funds to construct carefully graduated instruments that were put to use to make far more accurate measurements of the angular bearings of the heavenly bodies and of the angles these bodies subtend at points of observation on the Earth.

During the Alexandrian period Eratosthenes, Apollonius, Aristarchus, Hipparchus, Ptolemy, and dozens of other luminaries applied themselves to the study of geography and astronomy. By utilizing such resources the Alexandrians constructed the astronomical theory that was to dominate for over fifteen hundred years.

Outstanding among the contributions of the Alexandrians was the heliocentric hypothesis advanced by Aristarchus. History records little of the life of Aristarchus of Samos (c. 310–230 B.C.). We know him only through his work, which forms his sole biography. The fame of Aristarchus' heliocentric hypothesis has obscured his other achievements of durable significance. Early in his career the young astronomer made the first known attempt to calculate the sizes and distances of celestial objects. Although the stars and planets remained far too tiny and remote for human measurement, the steady accumulation of observations and the rapid progress of mathematics had put the sizes of the sun and the moon within reach of at least approximate calculation.

Aristarchus' calculations appear in the only one of his books that has come down to us, and by this fortunate circumstance we can trace the details of his thoughts. From a modern perspective his work on the

sizes and distances of the sun and moon is an exercise in simple trigonometry; however, Aristarchus' work predated the invention of trigonometry, and he had to be content with seeking upper and lower limits for such quantities rather than their precise values. His main
weapon is the splendid geometrical synthesis completed a generation
or two earlier (around the year 300 B.C.) by Euclid. Aristarchus accepts
Euclid and goes further, tacitly assuming some additional theorems of
his own. Then, ingeniously exploiting these new mathematical results,
he announces his three main conclusions. These deal with the distances of the sun and moon from the Earth and the ratios of the sizes
of the three bodies.

If we judge his results solely by comparison with modern figures,
we must pronounce Aristarchus' work a distinguished failure. The
fault lay not with Aristarchus' mathematics but with the observations
possible with the primitive instruments of his time. This gallant
attempt to measure the heavens with a few theorems of Euclid may
seem pathetic in the light of modern mathematics. But Aristarchus had
taken the first step along the main path of future progress. By asking
"How far?" and "How large?" he had begun the attack on two of the
central obstacles standing between humanity and a realistic picture of
the universe.

Nowhere in the sole surviving work does Aristarchus hint that the
Earth travels around the sun. A famous passage from the pen of
Archimedes, twenty-five years his junior, seems to leave no room for
doubt, however:

> Aristarchus of Samos brought out a book consisting of some hypotheses,
> [namely] that the fixed stars and the sun remain unmoved; that the Earth
> revolves about the sun in the circumference of a circle, the sun lying in
> the middle of the orbit.

We cannot now be sure of Aristarchus' motives. Heraclides had
given a potentially useful lead in teaching that the paths of Venus and
Mercury are centered on the sun; and Aristarchus' own reckoning of
sizes, combined with some intuition of the principles of dynamics,
may have convinced him that it would be physically more reasonable
to let the smaller body revolve around the larger. Alternatively, perhaps he regarded the concept of heliocentricity merely as an attractive
hypothesis, worth pursuing for its mathematical consequences. In any
case the idea proved too bold for its time and won little support. Moreover, the fact that Earth dwellers could not feel the rotation and rev-

olution of the Earth and the belief that the Earth was the natural center of the universe countered Aristarchus' scheme.

Soon after Aristarchus' pioneering attempt to measure the sizes and distances of the heavenly bodies, another brilliant scientist, working within the same tradition but setting his sights somewhat lower, announced the dimensions of an object that no one had ever seen in its entirety: the Earth.

Eratosthenes was born in or about the year 276 B.C. at Cyrene in northern Africa. Not content with successes in mathematics, astronomy, and geography, he ventured also into poetry, history, grammar, and literary criticism, and he earned the nickname "Beta" by ranking just below the best in all these pursuits; this was versatility, even for a Greek. As far as we can now say, he had few predecessors in his efforts to measure our globe, and they were crude.

Eratosthenes observed that at noon on the summer solstice the sun threw no shadow at Syene (Aswan), while at Alexandria, at the same hour, the pointer of a sundial threw a shadow amounting to 1/50 of a complete circle (Figure 19). Assuming that the two towns were 5000 stades (a Greek unit that we have been unable to calculate) apart on the same meridian, and that rays of sunlight arriving at different places on the Earth are parallel (a sophisticated idea for its time), Eratosthenes used a straightforward geometrical argument to show that the surface distance between Alexandria and Syene must be 1/50 of the Earth's circumference, which thus works out to 250,000 stades. He was wrong in two of his assumptions: (1) Alexandria and Syene are not in fact on the same meridian, and (2) Eratosthenes reckoned their

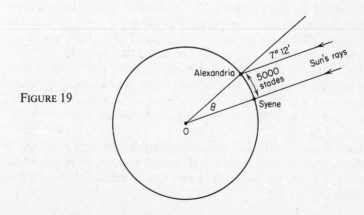

FIGURE 19

separation merely by the time required for the king's messengers to run between them. In any event, the short-term significance of Eratosthenes' achievement stemmed less from its degree of precision than from the example it set and the conviction it carried. Here was a continuation and encouragement of the growing trend toward quantitative cosmology, and another yardstick by which we might ultimately extend our measurements to the farthest reaches of the stars.

The quantitative methods of Aristarchus and Eratosthenes were soon extended to a quantitative theory of the solar system. Of course, these models of celestial motion—whether conceived merely as mathematical devices or as mirrors of physical reality—retained from first to last the ultimate goal of reproducing and predicting the paths actually traced by the bodies in the sky. The various modifications suggested by mathematical astronomers from the time of Eudoxus to the time of the thinkers whose work we are about to describe did take advantage of some of the ideas suggested by their predecessors.

The culmination and definitive achievement of Greek astronomy were the work of Hipparchus (died about 125 B.C.) and Claudius Ptolemy (died A.D. 168). Hipparchus lived most of his life in Rhodes. At the time he was active, about 150 B.C., Rhodes was a flourishing commercial and intellectual Greek state and a rival of Alexandria. Hipparchus knew fully the developments at Alexandria. He knew, for example, Eratosthenes' *Geographica* and wrote a criticism of it. He also had possession of older Babylonian observations and those made at Alexandria from about 300 to 150 B.C. Of course he made many observations of his own.

Hipparchus recognized that the scheme of Eudoxus, which supposed that the heavenly bodies were attached to rotating spheres with centers at the Earth's center, did not account for many facts observed by other Greeks and by Hipparchus himself. Instead of Eudoxus' scheme, Hipparchus supposed that a planet P (Figure 20) moved in a circle, the epicycle, at a constant speed, and that the center of this circle, Q, moved at constant speed on another circle with its center at the Earth. By properly selecting the radii of the two circles and the speeds of Q and P, he was able to arrive at an accurate description of the motion of many planets. The motion of a planet, according to this scheme, is like the motion of the moon according to *modern* astronomy. The moon revolves around the Earth at the same time that the Earth revolves around the sun. The motion of the moon around the sun is then like the motion of a planet around the Earth in Hipparchus' system.

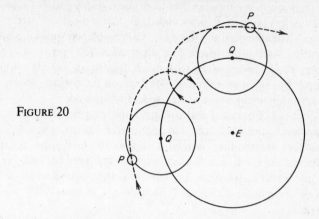

FIGURE 20

In the case of some heavenly bodies, Hipparchus found it necessary to use three or four circles, one moving on another. That is, a planet P moved in a circle about the mathematical point Q, while Q moved in a circle about the point R, and R moved about the Earth; again, each object or point traveled at its own constant speed. In some cases Hipparchus had to suppose that the center of the innermost circle or deferent was not at, but near, the center of the Earth. Motion in accordance with this latter geometrical construction was called eccentric, whereas when the center of the deferent was at the Earth, the motion was called epicyclic. By the use of both types and by the proper choice of radii and speeds of the circles involved, Hipparchus was able to describe quite well the motions of the moon, sun, and the five planets then known. With this theory an eclipse of the moon could be predicted to within an hour or two; solar eclipses were predicted less accurately.

We cannot survey all of Hipparchus' achievements, but we should note one splendid discovery that especially influenced later cosmology: the "precession of the equinoxes." The equinoctial points—the points of intersection of the *plane* of the celestial equator, the ecliptic, and the *plane* of the Earth's orbit—change slowly and regularly complete one period in about 26,000 years. Hipparchus reached his insight in the course of compiling the oldest of all star catalogues, which eventually gave the places of some 850 stars. He also estimated the length of the solar year to be 365 days, 5 hours, and 55 minutes, or about 6½ minutes too long.

It is worth mentioning that from the modern point of view Hipparchus was taking a step backward, for about a century earlier Aristarchus had advanced the theory that all the planets move around the sun. But observations made over a period of 150 years by the observatory at Alexandria along with older Babylonian records convinced Hipparchus of what we know today, that a heliocentric theory with planets, moving in *circles* about the sun will not do.

Instead of pursuing and perhaps improving on Aristarchus' idea, Hipparchus dismissed it as too speculative. Others rejected Aristarchus' idea because they deemed it impious to identify the corruptible matter of the Earth with the incorruptible heavenly bodies by regarding the Earth as a planet, a distinction quite solidly established in Greek thought and defended, though not dogmatically, even by Aristotle.

The second century after Christ witnessed the culmination of Greek cosmology. Its author was Claudius Ptolemaeus, whose birthplace was on the Nile. Like so many of the early heroes of our story, he has almost no biography; we are told only that he died at the age of seventy-eight, and that his astronomical observations, probably in Alexandria, spanned the years 127 to 151 A.D. In his own time, Ptolemy won renown no less for geography than for astronomy; he wrote also on optics, and he mingled some nonsense with his science by producing a volume on astrology. His enduring fame stems from his *Matematike Syntaxis,* or "Mathematical Composition." Arab translators called this book *al-megiste* ("the greatest"); hence derives the title *Almagest* under which it dominated the astronomy of Europe through fourteen hundred years.

Hipparchus' work is known to us because it survived in Ptolemy's *Almagest.* In its mathematical content the *Almagest* brought Greek trigonometry into the definitive form it was to retain for more than a thousand years. And in the field of astronomy it offered an original exposition of the geocentric theory of epicycles and eccentrics that has come to be known as the Ptolemaic theory. So accurate was it quantitatively and so long was it accepted that people were lured into regarding it as an absolute truth. This theory is the final Greek answer to Plato's problem of rationalizing the appearances of the heavens and is the first truly great scientific synthesis. With Ptolemy's completion of Hipparchus' work the evidence for mathematical design in the universe was complete to the "tenth" decimal place. However, Ptolemy, like Eudoxus, was explicit that his theory was only a mathematical construction.

Ptolemy was aware of Aristarchus' heliocentric theory but rejected it on the ground that the motion of an object is proportional to its mass. Hence if the Earth were moving it would leave behind lighter objects such as human beings and animals. His astronomy begins with the spherical shape of the heavens; this, he says, is humanity's oldest cosmological certainty. His own reasoning is based largely on observation, although echoes of the old a priori arguments persist: "The movement of the heavenly bodies ought to be the least impeded and most facile, [and] the circle among plane figures offers the easiest path of motion, and the sphere among solids." Ptolemy deems it necessary to give proofs—entirely observational in this case—that the Earth too is a sphere. As we have seen, he insists that our globe does not move, although he concedes that its rotation would produce some of the phenomena we see. The Earth is in the middle of the heavens; its size, says Ptolemy, continuing a tradition now well established, is a mere point compared to the distance of the stars.

Book III of the *Almagest* takes up the problem of the sun's path and gives essentially the solution found by Hipparchus, in which the center of the sun's motion is near but not at the Earth. "It would be more reasonable," Ptolemy says, "to stick to the hypothesis of eccentricity which is simpler and completely effected by one and not two movements." This telling passage reminds us that Ptolemy's deliberations were here guided by questions of elegance and economy, with no thought for the physical existence of these celestial circles. In the lunar theory Ptolemy finds that Hipparchus' model—epicycle on a deferent—fits the observations at the times of new and full moons, but breaks down at intermediate positions, where, as Hipparchus had seen, the apparent diameter of the epicycle seems to increase. Ptolemy therefore constructs a clever device that pulls the epicycle in toward the observer at the appropriate places. This adjusted model gives the moon's longitudes with high accuracy but has the serious drawback of implying wide variations in our satellite's distance from the Earth—variations wholly unconfirmed by any visible alteration in its apparent size.

Ptolemy next reckons the moon's distance from the Earth by comparing his observations with positions computed from his theory, obtaining a mean lunar distance of 29.5 Earth radii. Then he invokes the four-hundred-year-old argument of Aristarchus to deduce the remoteness of the sun, but here he goes badly astray, with an estimate less than half of Hipparchus' figure—which was ten times too low. No one bettered these estimates for the next fifteen hundred years. In

Books VII and VIII of the *Almagest* Ptolemy corrects and expands Hipparchus' catalogue of fixed stars, bringing the number of entries from 850 to 1022. He divides his stars into six classes, according to their respective "magnitudes". Nowadays this last term refers not to size but to apparent brightness; however, in antiquity all stars were thought to be equally remote from the Earth, so that brightness could be presumed proportional to size.

In Book IX Ptolemy launches his unique and crowning achievement: the first complete and rigorous account of the planets' erratic motions. His starting point is of course the unquestioned first axiom of celestial geometry, which he sets forth again:

> Our problem is to demonstrate, in the case of the five planets as in the case of the sun and moon, all their apparent irregularities as produced by means of regular and circular motions (for these are strangers to disparities and disorders).

Seldom in the history of science has any a priori principle governed people's thinking so completely and for so long.

As a first approximation, Ptolemy refers all planetary motions to the plane of the ecliptic, that is, the plane of the circular path of the sun, which he pictures as undergoing a slow rotation that produces the precession of the equinoxes. Furthermore, the simple epicycle-on-deferent scheme will not suffice for the planets, for it entails, contrary to observation, that the retrograde arcs are equal in length and uniformly spaced. Ptolemy counters this excessive symmetry by postulating an epicycle on an *eccentric*.

Within the fundamental eccentric–epicycle pattern the appearances can be saved, Ptolemy finds, only by postulating that each planet's epicycle moves uniformly not with respect to the deferent's center C but relative to another point, called the equant, Q in Figure 21. The Earth is at E, and $EC = CQ$. The planet moves around the epicycle in the same direction as the epicycle's center moves round the deferent (in contrast to the two opposite revolutions assumed in both the solar and lunar models); the retrogradations occur when the planet is on the side of the epicycle nearer to Earth. The single case of Mercury requires a further complication, similar to the device Ptolemy invented for the moon: the center of Mercury's deferent traces a small circle of its own, and by this means the little planet's epicycle is periodically drawn closer to Earth. An "inner" planet (Mercury or Venus) traces its epicycle in the planet's own "year" (which is equivalent to its time of revolution around the sun), while the epicycle's motion around the

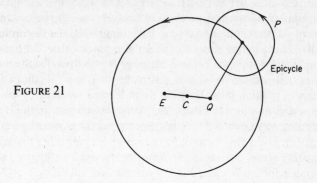

FIGURE 21

deferent takes one terrestrial year. The outer planets reverse the arrangement; here the time of motion of the epicycle on the eccentric equals what we now describe as the planet's own revolution around the sun, whereas the motion along the epicycle corresponds to what would be for us the time of Earth's revolution around the sun. Each epicycle is inclined to its deferent in such a way as to keep the epicycle roughly parallel to the ecliptic.

"Let no one," writes Ptolemy, "seeing the difficulty of our devices find troublesome such hypotheses," although the bewildered reader may wish to disagree. Science has, after all, no advance guarantee that nature will be simple.

From our point of view the equant is Ptolemy's masterstroke, a wholly original and completely successful scheme—and a foreshadowing of Kepler's ellipses. To some of the great astronomer's later critics, however, the intrusion of the equant seemed to compromise the sacred first principle of celestial motion, which insisted on uniformity relative only to the center of a circle. The equant became in some eyes a veritable scandal, which would help persuade Copernicus to move the Earth. From what strange causes do great revolutions spring!

To add to his brilliant schemes of the motions of the moon, sun, and planets, Ptolemy ordered them in distance from the Earth, although he made mistakes here, and he gave estimates of the sizes of the heavenly bodies that, as he knew, were crude because good astronomical instruments were not available.

Philosophical objections aside, the geometry of the *Almagest* was a triumph. However, we can readily believe that Ptolemy's inquiring mind would wish to supplement these avowedly imaginary circles with some account of the real material fabric of the heavens. The old sense of uniting mathematical theory and tangible reality, the descriptive work of the astronomer and the explanatory role of the physicist, had hardly abated during this early phase of geometrical model-building. Actually the distinction became progressively sharper, for the mathematics that insisted on noncircular paths for celestial bodies and on centers of rotation other than the Earth seemed painfully at variance with sound Aristotelian principles. Many cosmologists in the Hellenistic age simply ignored Aristotle's physics, but the growing complexity of mathematical devices must have evoked in many other thinkers an increasing sense of remoteness from reality, even perhaps a certain "nostalgia for the lost paradise" of Aristotelian simplicity.

Ptolemy has had in some quarters a bad "image." Some readers of the *Almagest* have been repelled by a real or fancied air of plodding pedantry and by the tortuous complications of a geometry that left the skies

> With centric and eccentric scribbled o'er,
> Cycle and epicycle, orb in orb.

However, within its self-imposed restriction to uniform circular motion, and excepting only its admission of excessively varying distances for the moon, Ptolemy's theory describes the celestial orbits with a precision fully matching the accuracy of the underlying observations; moreover, its multiplication of circles attests to the great astronomer's courage and skill in the face of nature's complexity. The introduction of the equant was mathematical creativity of the first order, and sets Ptolemy apart from even the most ingenious of his predecessors. The *Almagest* must be ranked among the most influential books in the history of science, although many features of his exposition, notably the "central" immobile Earth, retained the deepest convictions of everyday experience and the accumulated "wisdom" of centuries.

From the standpoint of the search for truths, it is noteworthy that Ptolemy, like Eudoxus, fully realized that his theory was just a convenient mathematical description that fit the observations and was not necessarily the true design of nature. For some planets he had a choice of alternative schemes, and he chose the mathematically simpler one. Ptolemy says in Book XIII of his *Almagest* that in astronomy one ought to seek as simple a mathematical model as possible. However, Ptolemy's mathematical model was received as the truth by the Christian world.

Ptolemaic theory offered the first reasonably complete evidence of the uniformity and invariability of nature and is the final Greek answer to Plato's problem of rationalizing the appearances. Overall, the great significance of Ptolemaic theory is that it demonstrated the power of mathematics to rationalize complex and even mysterious physical phenomena. The problem of understanding nature and even discovering totally unknown phenomena received impetus and encouragement from its first majestic success.

IV

The Heliocentric Theory of Copernicus and Kepler

Nevertheless it moves.   Galileo

The subject of this chapter is an oft-told tale concerning the adoption of a heliocentric theory of our planetary system, which replaced the Ptolemaic geocentric theory. Of course, the heliocentric theory seems now to be the correct one—but why should we accept it? It is contrary to our basic sensations. Does mathematics have anything to do with the acceptance of such a radical change in our conception of the physical world?

According to the heliocentric theory, the Earth rotates on its axis and completes a rotation every twenty-four hours of our measurement of time. What this means is that a person on the equator rotates through 25,000 miles in twenty-four hours or roughly at the rate of 1000 miles per hour. We may judge the incredible magnitude of this speed by our experience with speeds of 100 miles per hour in an automobile. Moreover, the Earth revolves around the sun at the rate of 18 miles per second or 64,800 miles per hour, another incredible speed. Yet here on Earth we do not feel either the rotational or the revolving motion. Moreover, if we are rotating at a speed of even 100 miles an hour, why are we not thrown off into space? Most of us have ridden on a merry-go-round, which rotates at about 100 feet per second, and felt the force that would eject us into space if we did not hold on to some fixed object on the merry-go-round.

Yet today we accept the heliocentric theory as factual, although traces of the older geocentric theory are still in our everyday language. We still say that the sun rises in the East and sets in the West. Accordingly, the sun moves, not we on a rotating Earth.

Why did mathematicians and astronomers make the drastic change to a heliocentric theory? As we shall see, the role of mathematics was decisive in this revolution. We have already seen (Chapter II) that the Europeans learned about the Greek works, which emphasized the mathematical design of nature. This belief was reinforced by the Catholic doctrine, which dominated in the Middle Ages, that God had designed the universe. Presumably mathematics was the essence of this design.

During the Italian Renaissance the Greek mathematical works were recovered from numerous sources and bought by the literate. Perhaps the enterprising merchants of the Italian towns received more than they bargained for when they aided the revival of Greek culture. They sought merely to promote a freer atmosphere; they reaped a whirlwind. Instead of continuing to dwell and prosper on firm ground, the terra firma of an immovable Earth, they found themselves clinging precariously to a rapidly spinning globe that was speeding about the sun at an inconceivable rate. It was probably sorry recompense to these merchants that the very same theory that shook the Earth free and set it spinning and revolving also freed the human mind.

The reviving Italian universities were the fertile soil for these new blossoms of thought. There, Nicolaus Copernicus became imbued with the Greek conviction that nature's behavior can be described by a harmonious medley of mathematical laws, and there, too, he became acquainted with the hypothesis—also Hellenic in origin—of planetary motion about a stationary sun. In Copernicus's mind these two ideas coalesced. Harmony in the universe demanded a heliocentric theory, and he became willing to move heaven and earth to establish it.

Copernicus was born in Poland in 1473. After studying mathematics and science at the University of Krakow, he decided to go to Bologna, where learning was more widespread. There he studied astronomy under the influential teacher Domenico Maria Novara, a foremost Pythagorean. In 1512 he assumed the position of canon of the cathedral of Frauenburg in East Prussia, his duties being that of steward of church properties and justice of the peace. During the remaining thirty-one years of his life he spent much time in a little tower of the cathedral closely observing the planets with the naked eye and making untold measurements with crude homemade instruments. The rest of his spare time he devoted to improving his new theory of the motions of heavenly bodies.

Copernicus's published works gave unmistakable, if indirect, indications of his reasons for devoting himself to astronomy. Judging by these, his intellectual and religious interests were dominant. He valued his theory of planetary motion, not because it improved navigational procedures but because it revealed the true harmony, symmetry, and design in the divine workshop. It was wonderful and overpowering evidence of God's presence. Writing of his achievement, which was thirty years in the making, Copernicus could not restrain his gratification: "We find, therefore, under this orderly arrangement a wonderful symmetry in the universe, and a definite relation of harmony in the motion and magnitude of the orbs, of a kind that is not possible to obtain in any other way." He does mention in the preface to his major work *On the Revolution of the Heavenly Spheres* (1543), that he was asked by the Lateran Council to help in reforming the calendar, which had become deranged over a period of many centuries. He writes that he kept this problem in mind, but it is quite clear that it never dominated his thinking.

By the time Copernicus tackled the problem of the motions of the planets, the Arabs, in their efforts to improve the accuracy of Ptolemaic theory, had added more epicycles, and their theory required a total of seventy-seven circles to describe the motion of the sun, moon, and the five planets then known. To many astronomers, including Copernicus, this theory was scandalously complex.

Harmony demanded a more pleasing theory than the complicated extension of Ptolemaic theory. Copernicus, having read that some Greek authors, notably Aristarchus, had suggested the possibility that the Earth revolved about a stationary sun and rotated on its axis at the same time, decided to explore this possibility. He was in a sense overimpressed with Greek thought, for he also believed that the motions of heavenly bodies must be circular or, at worst, a combination of circular motions, because circular motion was "natural" motion. Moreover, he also accepted the belief that each planet must move at a constant speed on its epicycle, and that the center of each epicycle must move at a constant speed on the circle that carried it. Such principles were axiomatic for him. Copernicus even added an argument that shows the somewhat mystical character of sixteenth-century thinking. He found that a variable speed could be caused only by a variable power; but God, the cause of all motions, is constant.

The upshot of his reasoning was that he used the scheme of deferent and epicycle to describe the motions of the heavenly bodies; however, the all-important difference in his version was that the sun was at the center of each deferent, while the Earth itself became a planet moving about the sun and rotating on its axis. Thus, he achieved considerable simplification.

To understand the changes Copernicus introduced, we shall confine ourselves to a simplified example. Copernicus observed that by having the planet *P* revolve about the sun *S* (Figure 22), and by having the Earth *E* also revolve about the sun, the positions of *P* as observed from *E* would still be the same. Hence the motion of the planet *P* is described by one circle, whereas the geocentric view calls for two circles. Of course, the motion of a planet around the sun is not strictly circular, and Copernicus added epicycles to the circles shown in Figure 22 to describe the motions of *P* and *E* more accurately. Nevertheless, he was able to reduce the number of circles required from seventy-seven to thirty-four to "explain the whole dance of the planets." Thus, the heliocentric view permitted a considerable simplification in the description of the planetary motions.

It is of interest to note that about 1530 Copernicus circulated a short account of his new ideas in a little tract called the *Commentariolus* and that Cardinal Nicolaus von Schönberg, the archbishop of Capua, wrote to Copernicus urging him to make the full work known and begging him for a copy to be made at the cardinal's expense. However, Copernicus dreaded the furor he knew his work would generate and for years shrank from publication. He entrusted his manuscript on the revolutions to Tiedemann Giese, bishop of Kulm, who, with the aid of Georg Joachim Rheticus, a professor at the University of Wittenberg, got the book published. A Lutheran theologian, Andreas Osiander, who assisted in the printing and feared trouble, added an unsigned preface of his own. Osiander stated that the new work was a hypothesis that allowed computations of the heavenly motions on geometrical principles, adding that this hypothesis was not the real situa-

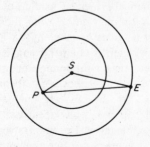

FIGURE 22

tion. Whoever takes for truth what was designed for different purposes, he added, will leave the science of astronomy a greater fool than when he approached it. Of course, Osiander did not reflect Copernicus's views, for Copernicus believed that the motion of the Earth was a physical reality. While lying paralyzed from a stroke, Copernicus received a copy of his book. It is unlikely that he was able to read it, for he never recovered. He died shortly afterward, in the year 1543.

The Copernican hypothesis of a stationary sun considerably simplified astronomical theory and calculations, but otherwise it was not impressively accurate. Copernicus fell far short of even 10 degrees of accuracy in predicting the angular positions of planets. He therefore tried variations on the basic plan of deferent and epicycle, with the sun, of course, always stationary and either at or near the center of the deferent. Although these variations were not much more successful, the failures did not diminish his enthusiasm for the heliocentric view.

When Copernicus surveyed the extraordinary mathematical simplification that the heliocentric hypothesis afforded, his satisfaction and enthusiasm were unbounded. He had found a simpler mathematical account of the motions of the heavens and hence one that must be preferred, for Copernicus, like all scientists of the Renaissance, was convinced that "Nature is pleased with simplicity, and affects not the pomp of superfluous causes." Copernicus could pride himself, too, on the fact that he had dared to think through what others, including Archimedes, had rejected as absurd.

As far as the mathematics of Copernican astronomy is concerned, it is purely a geometrical account involving no more than a reduction of a complex geometrical description to a simpler one. However, the religious and metaphysical principles affected by the change in theory were numerous and fundamental. On this account it is easy to see why a mathematician thinking only in terms of mathematics and unencumbered by nonmathematical principles would not hesitate to accept at once the Copernican simplification, while those who were guided chiefly or entirely by religious or metaphysical principles would not even venture to think in terms of a heliocentric theory. In fact, for a long time only mathematicians supported Copernicus.

As one would expect, a heliocentric theory that downgraded humanity's importance in the universe met severe condemnation. Martin Luther called Copernicus an "upstart astrologer" and "a fool who wishes to reverse the entire science of astronomy." John Calvin thundered, "Who will venture to place the authority of Copernicus above that of the Holy Spirit?" Do not the Scriptures say that Joshua commanded the sun and not the Earth to stand still? That the sun runs from one end of the heavens to the other? That the foundations of the Earth are fixed and cannot be moved?

The Inquisition condemned the new theory as "that false Pythagorean doctrine utterly contrary to the Holy Scriptures." The Catholic Church, in an official statement, called Copernicanism a heresy "more scandalous, more detestable, and more pernicious to Christianity than any contained in the books of Calvin, of Luther, and of all other heretics put together."

To these attacks Copernicus replied in a letter to Pope Paul III:

> If perhaps there are babblers who, although completely ignorant of mathematics, nevertheless take it upon themselves to pass judgment on mathematical questions and, improperly distorting some passages of the Scriptures to their purpose, dare to find fault with my system and censure it, I disregard them even to the extent of despising their judgment as uninformed.

Moreover, Copernicus added, the Bible may teach us how to go to heaven but not how the heavens go.

Although the hypothesis of a stationary sun simplified considerably astronomical theory and calculations, the epicyclic paths of the planets, as noted, did not quite fit observations. The definitive improvement was made some fifty years later by that almost incredible mystic, rationalist, and empiricist, Johann Kepler (1571–1630), a German who combined wonderful imaginative power and emotional exuberance with infinite patience in acquiring data and with meticulous adherence to the dictates of facts. The personal life of Kepler contrasts sharply with that of Copernicus. The latter had obtained an excellent education as a youth and lived a retired and secure life, which he was able to devote almost solely to his theorizing. Kepler, born in 1571 with delicate health, was neglected by his parents and received a rather poor education. Like most boys of his time who showed some interest in learning, he was expected to study for the ministry. In 1589 he enrolled at the German University of Tübingen where he learned astronomy from an enthusiastic Copernican, Michael Mästlin. Kepler was impressed by the new theory, but the superiors of the Lutheran Church were not and questioned Kepler's devoutness. Kepler's objections to the narrowness of the current Lutheran thought led him to abandon a ministerial career and accept the position of professor of mathematics and morals at the University of Graz in Styria, Austria, where he lectured on mathematics, astronomy, rhetoric, and Virgil. He was also called on to make astrological predictions, which at the time he seems to have believed in. He set out to master the art and practiced by checking predictions of his own fortunes. Later in life he became less credulous and used to warn his clients, "What I say will or will not come to pass."

At Graz, Kepler introduced the new calendar advocated by Pope Gregory XIII. The Protestants rejected it, because they preferred to be at variance with the sun rather than in accordance with the Pope. Unfortunately, the liberal Catholic ruler of Styria was succeeded by an intolerant one, and Kepler found life uncomfortable. Although he was protected by the Jesuits for a while, and could have stayed by professing Catholicism, he refused to do so and finally left Graz.

In 1600 he secured a position as assistant to the famous astronomical observer Tycho Brahe (1546–1601), who had been making the first large revision of astronomical data since Greek times. On Brahe's death in 1601 Kepler succeeded him as "Imperial Mathematician" to Emperor Rudolph II of Bohemia. This employer, too, expected Kepler to cast horoscopes for members of the court. Kepler resigned himself to this duty with the philosophical view that nature provided all animals with a means of existence. He was wont to refer to astrology as the daughter of astronomy who nursed her own mother.

About ten years after Kepler joined him in Prague, the Emperor Rudolph began to experience political troubles and could not afford to pay Kepler's salary; thus Kepler had to find another job. In 1612 he accepted the position of provincial mathematician at Linz, but other difficulties still plagued him. While he was at Prague Kepler's wife and one son had died. He remarried, but at Linz two more of his children died; in addition to personal tragedy, the Protestants did not accept him, and money was scarce, forcing him to struggle for existence. In 1620 Linz was conquered by the Catholic Duke Maximilian of Bavaria, and Kepler suffered still stronger persecution. Ill health began to weaken him. The last few years of his life were spent in trying to secure publication of more books, collecting salary owed to him, and searching for a new position.

Kepler's scientific reasoning is fascinating. Like Copernicus he was a mystic, and like Copernicus he believed that the world was designed by God in accordance with some simple and beautiful mathematical plan. "Thus God himself was too kind to remain idle, and began to play the game of signatures, signifying his likeness into the world; therefore I chance to think that all nature and the graceful sky are symbolized in the art of geometry." He says further in his *Mystery of the Cosmos* (1596), the mathematical harmonies in the mind of the Creator furnish the cause "why the number, the size, and the motion of the orbits are as they are and not otherwise." This belief dominated all his thinking.

Still, Kepler also had qualities that we now associate with scientists. He could be coldly rational. Although his fertile imagination triggered the conception of new theoretical systems, he knew that theories must fit observations and, in his later years, saw even more clearly that empirical data may indeed suggest the fundamental principles of science. He therefore sacrificed his most beloved mathematical hypotheses when he saw that they did not fit observational data, and it was precisely this incredible persistence in refusing to tolerate discrepancies, which any other scientist of his day would have disregarded, that led him to espouse radical ideas. He also had the humility, patience, and energy that enable great men to perform extraordinary labor.

During the years he spent as astronomer to the Emperor Rudolph, Kepler did his most serious work. Moved by the beauty and harmonious relations of the Copernican system, he decided to devote himself to the search for whatever additional geometrical harmonies the data supplied by Tycho Brahe's observations might suggest and, beyond that, to find the mathematical relations binding all the phenomena of nature to each other. His predilection for fitting the universe into a preconceived mathematical pattern, however, led him to spend years in following up false trails. In the preface to his *Mystery of the Cosmos,* we find him writing:

> I undertake to prove that God, in creating the universe and regulating the order of the cosmos, had in view the five regular bodies of geometry as known since the days of Pythagoras and Plato, and that he has fixed according to those dimensions, the number of heavens, their proportions, and the relations of their movements.

He therefore postulated that the radii of the orbits of the six planets were the radii of spheres related to the five regular solids in the following way. The largest radius was that of the orbit of Saturn. In a sphere of this radius he supposed a cube to be inscribed. In this cube a sphere was inscribed whose radius should be that of the orbit of Jupiter. In this sphere he supposed a tetrahedron to be inscribed and in this, in turn, another sphere whose radius was to be that of the orbit of Mars, and so on through the five regular solids. This allowed for six spheres, just enough for the number of planets known then (Figure 23).

FIGURE 23

Kepler soon realized that his beautiful theory was not accurate. Although his calculated interplanetary distances were very close to reality, they did not match exactly the spaces between the spheres separating the nested polyhedra.

Thus far, Kepler's work was subject to the same criticism that Aristotle used to attack the Pythagoreans: "They do not with regard to the phenomena seek for their reasons and cause but forcibly make the phenomena fit their opinions and preconceived notions and attempt to reconstruct the universe." However, the enlightened Kepler had too much regard for facts to persist with theories that failed to agree with observations and that could not yield accurate predictions.

It was after Kepler had acquired Brahe's data and had made additional observations of his own that he became convinced he must discard the astronomical patterns his predecessors, Ptolemy and Copernicus, and he himself had conceived. The search for laws that would fit these new data culminated in three famous results. The first two Kepler presented to the world in his book, *On the Motion of the Planet Mars,* published in 1609.

The first of these laws broke with all tradition and introduced the ellipse in astronomy. This curve had already been studied intensively by the Greeks about two thousand years earlier, and therefore, its mathematical properties were known. Whereas the circle is defined as the set of all points that are at a constant distance (the radius) from a fixed point, the ellipse can be defined as the set of all points the *sum* of whose distances from two fixed points is constant. Thus, if F_1 and F_2 are the fixed points (Figure 24), and P is a typical point on the ellipse, the sum $PF_1 + PF_2$ is the same no matter where P may be on the ellipse. The two fixed points F_1 and F_2 are called the foci. Kepler's first law states that each planet moves along an ellipse and that the sun is at one of the foci. The other focus is just a mathematical point at which nothing physical exists. Of course, each planet moves on its own ellipse, of which one focus, the one at which the sun exists, is the same for all the planets. Thus, the upshot of fifteen hundred years of attempts to use combinations of circles to represent the motion of each planet was the replacement of each combination by a simple ellipse.

Kepler's first law tells us the path pursued by a planet, but it does not tell us how fast the planet moves along this path; if we should observe a planet's position at any particular time, we would still not know when it will be at any other point on that course. One might expect that each planet would move at a constant velocity along its path, but observations, the final authority, convinced Kepler that this was not the case. Kepler's second discovery states that the area swept out by the line from the sun to the planet is constant. That is, if the planet moves from P to Q (Figure 25) in one month, say, and from P' to Q' in the same time, then the *areas F_1PQ and $F_1P'Q'$* are the same. Kepler was overjoyed to find that there was a simple way to state the

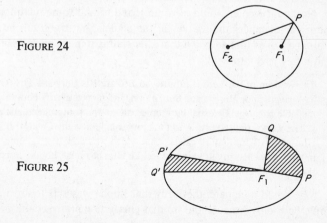

FIGURE 24

FIGURE 25

mathematical law of planetary velocities. Apparently God preferred constant area to constant speed.

Another major problem remained unsolved. What law described the distances of the planets from the sun? The problem is complicated by the fact that a planet's distance from the sun is not constant, and Kepler searched for a new principle that would take this into account. He believed that nature was not only mathematically but harmoniously designed. Thus, he believed that there was a music of the spheres that produced a harmonious tonal effect, not one given off in actual sounds but discernible by some translation of the facts about planetary motions into musical notes. He followed this lead, and after an amazing combination of mathematical and musical arguments, arrived at the law that if T is the period of revolution of any planet and D is its mean (or average) distance from the sun then

$$T^2 = kD^3$$

where k is a constant that is the same for all the planets. (Note that the correct value for D is actually the semimajor axis of each elliptical path.) This statement is Kepler's third law of planetary motion and the one that he triumphantly announced in his book *The Harmony of the World* (1619).

Because the Earth's mean distance from the sun is 93 million miles and the time of revolution is one year, we can substitute these values of D and T in the law and determine k. Then the law can be used to compute a planet's mean distance from the sun if one knows its period of revolution, or vice versa.

No doubt Kepler would have preferred to find some relationship among the distances themselves of the planets, but the result he did obtain overjoyed him so much that after stating it in his book he broke forth into a paean to God:

> The wisdom of the Lord is infinite; so also are His glory and His power. Ye heavens, sing His praises! Sun, moon, and planets glorify Him in your ineffable language! Celestial harmonies, all ye who comprehend His marvelous works, praise Him. And thou, my soul, praise thy Creator! It is by Him and in Him that all exists. That which we know best is comprised in Him, as well as in our vain science. To Him be praise, honor, and glory throughout eternity.

We should note parenthetically that Kepler was able to formulate simple laws only because the mutual gravitational attractions of the planets are relatively small and because the mass of the sun happens to be much larger than the masses of the planets. Nevertheless, Kepler's work was a major innovation and a great advance in the heliocentric theory.

Because we are today taught to accept the heliocentric theory and Kepler's laws, we no longer appreciate the full significance of Copernicus's and Kepler's achievements. It will profit us to go back for a moment to survey the setting in which these men worked and to see what their mathematics really accomplished.

We should recall that Copernicus and Kepler worked in the sixteenth and seventeenth centuries. The geocentric view had been in force since the days of Ptolemy and fitted in very neatly with the thoroughly entrenched religious doctrines. The Earth was at the center of the universe, and humans were the central characters in the universe. For us the sun, moon, and stars had been specially created. The heliocentric theory denied this basic dogma, implying instead that humanity was an insignificant speck of dust on one of many whirling globes. Was it likely then that humanity was the chief object of God's ministrations? The new astronomy also destroyed Heaven and Hell, which had very reasonable geographical locations in the geocentric view.

Copernicus and Kepler were highly religious, and yet both denied one of the central doctrines of Christianity. By displacing the Earth, Copernicus and Kepler removed a cornerstone of Catholic theology and imperiled the whole structure. Copernicus attacked the argument that the Earth is at the center of the universe by pointing out that the size of the universe is so immense compared to the Earth that to speak of a center is meaningless. However, this argument put him all the more into opposition with the Church.

There were also very reasonable scientific objections to a heliocentric theory. If the Earth is in motion, the directions of the stars, which seem to be fixed in the heavens, should be different at different points of the Earth's orbit. However, this difference was not observed in the sixteenth and seventeenth centuries. Copernicus disposed of this objection by stating that the distance of the stars is immense in comparison with the orbit of the Earth. However, his opponents argued that the distances necessary to make the change in direction insensible were inconsistent with the fact that the stars were clearly observable.

Copernicus's explanation proved to be the correct one, although even he would have been astonished to learn some of the modern figures for the distances of the stars from the Earth. The change in direction of the stars when viewed from one position on the Earth's orbit as opposed to another was first measured by the mathematician Friedrich Wilhelm Bessel (1784–1846) in 1838 and proved to be 0.76 seconds of arc (0.76″) for the nearest star.

The above objection was taken seriously by only a few specialists, but there were other valid scientific objections to a moving Earth that every layperson appreciated. Neither Copernicus nor Kepler could explain how the heavy matter of the Earth could be put into and kept in motion. That the other planets were also in motion even under the geocentric view did not disturb people who believed the heavenly bodies were composed of light material and therefore could readily move. The best answer that Copernicus could offer was that it was natural for any sphere to move. Equally troublesome was the question: Why did not the Earth's rotation cause objects on it to fly off into space, just as an object on a rotating platform flies off? Indeed, Ptolemy had rejected the rotation of the Earth for this very reason. Furthermore, why did not the Earth itself fly apart? To the latter question Copernicus replied that because motion was natural it could not have the effect of destroying the body. He also countered by asking why the skies did not fall apart under the very rapid daily motion presumed by the geocentric hypothesis. Entirely unanswered was the objection that, if the Earth rotated from west to east, an object thrown up into the air should fall back to earth west of its original position. Again, if, as practically all scientists since Greek times believed, the motion of an object was proportional to its weight, why did not the Earth leave behind objects of lesser weights? Even the air surrounding the Earth should have been left behind. Although he could not account for the fact that all objects on the Earth moved with it, Copernicus "explained" the continued presence of the atmosphere by arguing that the air was earthy and so rotated in sympathy with the Earth. Kepler advanced the theory that an object thrown upward returns to its starting point while the Earth rotates under it, because magnetic invisible chains attach the object to the Earth.

Another and most reasonable argument against the new heliocentric theory was that nobody could *feel* either the rotation or the revolution of the Earth. On the other hand, everyone apparently did see the motion of the sun. To the famous astronomer Tycho Brahe, these and other arguments were conclusive proof that the Earth must be stationary.

The substance of all these arguments is that a rotating and revolving Earth did not fit in with Aristotle's physical theory of motion, which was commonly accepted in Copernicus's and Kepler's times. What was needed was a totally new theory of motion.

Against all these arguments, Copernicus and Kepler had one masterly retort. Each had achieved mathematical simplification and an overwhelmingly harmonious and aesthetically superior theory. If mathematical relationships are the goal of scientific work, and if a better mathematical account could be given, then this fact, reinforced by the belief that God had designed the world and would clearly have used the superior theory, was sufficient to outweigh all objections. Each believed and clearly stated that his work revealed the harmony, symmetry, and design of the divine workshop and overpowering evidence of God's presence.

In view of the variety of weighty objections to the heliocentric theory, Copernicus's and Kepler's willingness to pursue it is one of the enigmas of history. Almost every major intellectual creation is preceded by decades and even centuries of ground-breaking work that, in retrospect at least, makes the decisive step appear to be natural. Yet Copernicus had no immediate scientific predecessors, and his sudden adoption of the heliocentric view, despite the unquestioned acceptance for fifteen hundred years of the geocentric view, today seems decidedly unnatural. In the company of the astronomers of the sixteenth century Copernicus stands forth as a colossus.

It is true, as we have noted, that Copernicus had read the Greek works in which several astronomers had advanced the idea that the Earth was in motion, but none had attempted to work out a mathematical theory on this basis, whereas the geocentric theory had been intensively developed. Nor did Copernicus's observations suggest that something radically new was called for. His instruments were as crude as those of his predecessors and his observations no better. He was disturbed by the complexity of Ptolemaic theory, which by his time had become entangled in many more epicycles to make the theory fit Arabian and European observations. In the magnificent dedication of his book to Pope Paul III, Copernicus remarks that he was first induced to seek a new theory when he found that mathematicians were arguing among themselves as to the soundness of Ptolemaic theory. Nevertheless, historically, the appearance of his work is as surprising as a mountain suddenly rising from a calm sea.

Actually, religious convictions of a special sort account for the direction of the work of Copernicus and Kepler. The glimpse of a new possibility that might reveal the greater grandeur of God was sufficient to arouse them and to fire their thoughts. The results of their efforts satisfied their expectations of harmony, symmetry, and design in the divine workshop. The mathematical simplicity of the new theory was proof that God would have used it in preference to a more complicated design.

Ptolemy had asserted that in explaining phenomena it is necessary to adopt the simplest hypothesis that will fit the facts. Copernicus turned this very argument against Ptolemaic theory, and because he believed the universe to be the work of God, he interpreted the simplicity he had found to be the true design. Because Kepler's mathematics was even simpler, he had all the more reason to believe that he had found the very laws that God had incorporated in the construction of the universe. Kepler said of his theory, "I have attested it as true in my deepest soul and I contemplate its beauty with incredible and ravishing delight."

There was also a mystical element in their thinking that now seems anomalous in great scientists. The inspiration to conceive and carry out a heliocentric theory came from some vague and even primitive response to the power of the sun. Copernicus wrote: "The earth conceives from the sun and the sun rules the family of stars." He reinforced this argument by the statement, "For who could in this most beautiful temple place this lamp in another or better place than that from which at the same time it can illuminate the whole."

Despite the mystical and religious influences, however, Copernicus and Kepler were thoroughly rational in rejecting any speculations or conjectures that did not agree with observations. What distinguishes their work from medieval vaporizings is not only the mathematical framework of their theories but also their insistence on making the mathematics fit reality. In addition, the preference they showed for a simpler mathematical theory is a thoroughly modern scientific attitude.

Despite the weighty scientific arguments against a moving Earth, despite the religious and philosophical conservatism, and despite the affront to common sense, the new theory gradually won acceptance. Mathematicians and astronomers were impressed, especially after Kepler's work, with the simplicity of the new theory. Also, it was far more convenient for navigational computations and calendar reckoning, and hence many geographers and astronomers who were not convinced of its truth began to use it nevertheless.

It is not surprising that at first only mathematicians supported the new theory. Only a mathematician and only one convinced that the universe was mathematically and simply designed would have had the mental fortitude to disregard the prevailing philosophical, religious, and physical beliefs and to work out the mathematics of a revolutionary astronomy. Only one possessed of unshakable convictions as to the importance of mathematics in the design of the universe would have dared to affirm the new theory against the mass of powerful opposition it was sure to encounter.

The heliocentric theory found an extremely able defender in Galileo Galilei (1564–1642). Born in Florence, he entered the University of Pisa at the age of seventeen to study medicine. His reading of Euclid and Archimedes fired his interest in mathematics and science, and so he turned to these fields.

An offer of a professorship at the University of Padua prompted Galileo's move to northeastern Italy in 1592. Padua was then within the realm of the progressive Venetian Republic, and Galileo enjoyed total academic freedom. In 1610 the Grand Duke Cosimo de' Medici, his former pupil, engaged Galileo as court philosopher and mathematician. The move to Florence marked the end of his teaching duties and the beginning of his career as a full-time scientist.

In the summer of 1609 Galileo caught wind of a Dutch invention by means of which distant objects were distinctly seen as if nearby. Galileo wasted no time in constructing his own telescope and gradually improving the lenses until he reached a magnification of thirty-three times. In a dramatic demonstration to the Venetian Senate, Galileo showed the telescope's power to sight approaching hostile warships nearly two hours before their arrival.

But Galileo had grander plans for his instrument. Fixing his telescope on the moon, he observed vast craters and imposing mountains, thereby destroying the notion of a smooth lunar surface. By viewing the sun he discovered mysterious spots on its surface. He discovered also that Jupiter possesses four orbiting moons. (We can now observe sixteen.) This discovery showed that a planet, like the Earth, can have satellites. Galileo announced this discovery in *The Starry Messenger* (1610), describing the four moons of Jupiter that he observed in 1610 as "bodies that belong not to the inconspicuous multitude of fixed stars, but to the bright ranks of planets." In a rare act of political astuteness, he dubbed these moons the "Medici Planets," to honor his powerful Florentine patron.

Copernicus had predicted that if human sight could be enhanced, then we would be able to observe phases of Venus and Mercury, that is, to observe that more or less of each planet's hemisphere facing Earth is lit up by the sun, just as the naked eye can discern phases of the moon. Galileo did discover the phases of Venus. Here was further evidence that the planets were like Earth and certainly not perfect bodies composed of some special ethereal substance, as Greek and medieval thinkers had believed. The Milky Way, which had hitherto appeared to be just a broad band of light, could be seen with the telescope to be composed of thousands of stars, each of which gave off light. Thus, there were other suns and presumably other planetary systems suspended in the heavens; moreover, the heavens clearly contained more than seven moving bodies, a number that had been accepted as sacrosanct. His observations convinced him that the Copernican system was correct.

Annoyed by Galileo's defense of the heliocentric doctrine, the Roman Inquisition in 1616 declared the doctrine heretical and censored it, and in 1620 the Inquisition forbade all publications teaching this doctrine. Despite the earlier ecclesiastical prohibition of works on Copernicanism, Pope Urban VIII did give Galileo permission to publish a book on the subject, for the Pope believed there was no danger that anyone would ever prove the new theory necessarily true. Accordingly, in his *Dialogue on the Great World Systems* (1632) Galileo compared the geocentric and heliocentric doctrines. With the aim of pleasing the Church and so passing the censors, he incorporated a preface to the effect that the latter theory was only a product of the imagination. Galileo had been admonished to present the two theories, geocentric and heliocentric, as equally valid, but the bias in favor of the latter was evident. Unfortunately, Galileo wrote too well, and the Pope began to fear that the argument for heliocentrism, like a live bomb wrapped in silver foil, could still do a great deal of damage to the Catholic faith. Galileo was again called by the Roman Inquisition and compelled on the threat of torture to declare: "The falsity of the Copernican system cannot be doubted, especially by us Catholics." In 1633 Galileo's *Dialogue* was put on the Index of Prohibited Books, a ban that was not lifted until 1822.

We who live in an age of space exploration, with spaceships that can take people to the moon and travel to the farthest planets of our solar system, no longer can doubt the truth of the heliocentric theory. However, the people of the seventeenth and eighteenth centuries, even those able to understand the writings of Copernicus, Kepler, and Galileo, could with good reason be skeptical. The evidence of the senses went against the theory, and the mathematical arguments of Copernicus and Kepler, which apart from philosophical beliefs rested on the relative simplicity of the heliocentric theory, carried little weight.

There is one more implication that modern science has perceived in the work of Copernicus and Kepler. The same observational data that Hipparchus and Ptolemy organized in their geocentric theory of deferent and epicycle can also be organized under the heliocentric theory of Copernicus and Kepler. Despite the belief of the latter that the new theory was true, the modern view is that either theory will do and there is no need to adopt the heliocentric hypothesis except to gain mathematical simplicity. Reality seems far less knowable than Copernicus and Kepler believed, and today scientific theories are regarded as human inventions. Modern astronomers might agree with Kepler that the heavens declare the glory of God and the firmament showeth His handiwork; however, they now recognize that the mathematical interpretations of the works of God are their own creations, and mathematical simplicity wins out despite their sensations. How then do we determine what is real in our physical world?

V

Mathematics Dominates Physical Science

> *So that we may say the door is now opened, for the first time, to a new method fraught with numerous and wonderful results which in future years will command the attention of other minds.*
> Galileo Galilei

In pursuit of our major theme of how mathematics reveals and determines our knowledge of the physical world, we have just seen that for mathematical reasons people accepted the heliocentric theory of planetary motions. Whether this theory would have survived were the mathematical advantages not clear is questionable, especially in view of the opposition of the Church. The theory did survive, however, as well as many others that, as we shall see, either defy our sense perceptions or compel us to accept physical realities not at all accessible to sense perceptions. The basic reason for the acceptance of such theories is the rise, beginning in the seventeenth century, of the domination of science by mathematics coupled with the belief common to that era that mathematics is truth. We shall diverge here from our main purpose of discovering what mathematics discloses about our physical world to see how mathematics came to be the acme of truth and a superbly effective instrument for the study of the physical world.

Sir Isaac Newton once said that he stood on the shoulders of giants. The tallest of these giants were René Descartes (1596–1650) and Galileo Galilei (1564–1642). The great achievements of modern mathematics are due not only to an increased emphasis on mathematics but to a new methodology initiated and pursued by these two outstanding seventeenth-century thinkers. Let us consider their contributions in turn.

The established system of scientific thought and the nature of scientific activity before the seventeenth century derived from Aristotle, and the chief characteristic of the Aristotelian approach to nature was the search for material or qualitative explanations. The Aristotelians tried to explain phenomena on the Earth in terms of qualities or substances they had come to believe were basic—for example, hot and cold, wet and dry. Combinations of these qualities were thought to give rise to the four elements—earth, air, fire, and water. Thus, hot and dry qualities produced fire, hot and wet produced air, and so on. Each of the four elements had a characteristic motion. Fire, the lightest, naturally sought the heavens, while earthy matter naturally sought the center of the Earth. Aristotle also treated what he called violent motions, in which one body hits and propels another.

Solids, fluids, and gases were regarded as three different kinds of substances possessing different fundamental qualities—rather than, as we say, different states of the same substance. The transition from fluid to gas meant to the Greeks the loss of one quality and the acquisition of another. Different objects differed in fundamental qualities. For example, it was thought that in changing mercury into gold one took from mercury the quality that contributed fluidity and substituted the quality that possessed rigidity. This idea of basic qualities was still pursued during the early stages of modern chemistry. Thus, sulfur possessed the substance of combustibility, called phlogiston; salt, the substance of solubility; and metals, the basic substance of mercury. Heat, until the nineteenth century, was a substance called caloric, which could be gained or lost by bodies as they acquired or lost heat.

The Aristotelians sought to classify objects according to the qualities or basic substances they contained; hence one of their major goals was classification, a method still basic in biology. To explain how one event brought about another, the Aristotelians built an elaborate scheme in which all phenomena came about because there were four types of causation: the material cause, the formal cause, the effective cause, and the final cause. Just to distinguish among these causes, let us consider an artist making a statue. The material cause would be the stone and the artist's tools; the formal cause would be the design the artist had in mind; the effective cause would be the artist actually chipping away at the stone; and the final cause would be the purpose the statue would serve in beautifying some room or building. The final or teleological cause was the most important, because it gave meaning to the entire activity. Where was mathematics in this scheme? Because mathematics to the Greeks was largely geometry, and geometry dealt with figures, mathematics was of use mainly in describing the formal cause—a rather limited role.

For several reasons the Aristotelian approach to nature dominated in the medieval period and the Renaissance. Aristotle's writings were comprehensive, and they were more widely disseminated than those of the other Greeks. Furthermore, Aristotle's theory of final cause had been adopted and espoused by Catholic theology. The explanation of human life on the Earth was that it prepared us for Heaven, and quite generally the Church explained earthly phenomena as serving purposes intended by God.

Although we need not trace the history of the Renaissance here, we can surely say without question that by 1600 the European scientists were unquestionably impressed with the importance of mathematics for the study of nature. The strongest evidence of this conviction was the willingness of Copernicus and Kepler to overturn astronomy, mechanics, and religious doctrines for the sake of a theory that in their time had only mathematical advantages.

Why did the scientific activity that was initiated in the seventeenth century prove so effective? Were the contributors such as Descartes, Galileo, Newton, Huygens, and Leibniz greater intellects than those found in earlier civilizations? Hardly. Was it because of the increased use of observation, experiment, and induction—methods urged by Roger Bacon and Francis Bacon? Apparently not. The turn to observation and experimentation may have been an innovation in the Renaissance, but it was a method of approach at least familiar to Greek scientists. Nor does the mere use of mathematics in scientific studies explain the amazing accomplishments of modern science, for although the seventeenth-century scientist knew that the goal of his work should be to ferret out the mathematical relationships behind various phenomena, the search for such relationships in nature was not new to science.

The astonishing successes of modern science and the enormous impetus to create new mathematics derived from the science of the seventeenth and later centuries probably would not have come from following in the footsteps of the past. In the seventeenth century Descartes and Galileo reformed and reformulated the very nature of scientific activity. They selected the concepts that science should employ, redefined the goals of scientific activity, and altered the very methodology of science. Their reformulation not only imparted unprecedented power to science but also bound science indissolubly to mathematics. In fact, their plan practically reduced theoretical science to mathematics. To understand the spirit that moved mathematics from the seventeenth century onward, we must first examine the ideas of Descartes.

When Descartes was still in school at La Flèche, he began to ponder how it was that humanity professed to know so many truths. Partly because he had a critical mind and partly because he lived at a time when the world outlook that had dominated Europe for a thousand years was being vigorously challenged, Descartes could not be satisfied with the tenets so forcibly and so dogmatically pronounced by his teachers and by leaders of sects. He felt all the more justified in his doubts when he realized that he was in one of the most celebrated schools of Europe and that he was not an inferior student. At the end of his course of study he concluded that there was no sure body of knowledge anywhere. All his education had advanced him only to the point of discovering humanity's ignorance.

To be sure, he did recognize some values in the usual type of studies. He agreed that "eloquence has incomparable force and beauty; that poesy has its ravishing graces and delights"; however, he judged these to be gifts of nature rather than fruits of study. He revered theology because it pointed out the path to heaven, and he too aspired to heaven, but, "being given assuredly to understand that the way is not less open to the most ignorant than to the most learned, and that the revealed truths which lead to heaven are above our comprehension," he did not presume to subject them to the impotence of his reason. Philosophy, he granted, "affords the means of discoursing with an appearance of truth on all matters, and commands the admiration of the more simple." Yet it had produced nothing (thus far) that was beyond dispute or above all doubt. After criticizing other studies, including jurisprudence, medicine, and morality, Descartes found only mathematics to be the sure road to truth.

Descartes was explicit in his belief that mathematics was the essence of science. He wrote that he "neither admits nor hopes for any principles in Physics other than those which are in Geometry or in abstract Mathematics, because thus all phenomena of nature are explained and some demonstrations of them can be given." The objective world is space solidified or geometry incarnate, and therefore its properties should be deducible from the first principles of geometry.

Descartes elaborated on why the world must be accessible to mathematics. He insisted that the most fundamental and reliable properties of matter are shape, extension in space, and motion in space and time. Because shape is just a matter of extension in space, extension and motion are the basic realities. Thus Descartes asserts; "Give me extension and motion and I shall construct the universe."

To establish truths, said Descartes, the method of mathematics should be employed because the method transcends its subject matter: "It is a more powerful instrument of knowledge than any other that has been bequeathed to us by human agency, as being the source of all others." In the same vein he continues:

> All the sciences which have for their end investigations concerning order and measure are related to mathematics, it being of small importance whether this measure be sought in numbers, forms, stars, sounds, or any other object; that accordingly, there ought to exist a general science which should explain all that can be known about order and measure, considered independently of any application to a particular subject and that, indeed, this science has its own proper name consecrated, by long usage, to wit, mathematics. And a proof that it far surpasses in facility and importance the sciences which depend upon it is that it embraces at once all the objects to which these are devoted and a great many others besides.

And so he concludes that

> The long chains of simple and easy reasonings by means of which geometers are accustomed to reach the conclusions of their most difficult demonstrations had led me to imagine that all things to the knowledge of which man is competent are mutually connected in the same way.

From his study of mathematical method he isolated in his *Rules for the Direction of the Mind* the following principles for securing exact knowledge in any field. He would accept nothing as true that was not so clearly and distinctly so to his own mind as to exclude all doubt. He would divide difficulties into smaller ones; he would proceed from the simple to the complex; and finally, he would enumerate and review the steps of his reasoning so thoroughly that nothing would be omitted. The mind's immediate apprehension of basic, clear, and distinct truths, this intuitive power, and the deduction of consequences are the essence of his philosophy of knowledge. How then does he differentiate between acceptable and unacceptable intuitions? The answer lies in the words "clear and distinct." In the third of his *Rules* he states:

> Concerning the objects we propose to study, we should investigate not what others have thought nor what we ourselves conjecture, but what we can intuit clearly and evidently or deduce with certainty, for there is no other way to acquire knowledge.

There are, then, according to Descartes, only two mental acts that enable us to arrive at knowledge without any fear of error: intuition and deduction. These he proceeds to define in the passage that follows, another example of how the *Rules* are virtually indispensable for any clear understanding of his method:

> By *intuition* I understand not the unstable testimony of the senses, nor the deceptive judgment of the imagination with its useless constructions; but a conception of a pure and attentive mind so easy and so distinct that no doubt at all remains about that which we are understanding. Or, what amounts to the same thing, intuition is the undoubting conception of a pure and attentive mind, which comes from the light of reason alone and is more certain even than deduction because it is simpler; although, as we have noted above, the human mind cannot err in deduction, either. Thus everyone can see by intuition that he exists, that he thinks, that a triangle is bounded by only three lines, a sphere by a single surface, and other similar facts.

Again, to reinforce his conviction that humanity can through mathematics discover the laws of nature, he argues in his *Discourse on the Method of Rightly Conducting the Reason and Seeking Truth in the Sciences* (1637) that because God would not deceive us, we can be sure that the truths that the mind recognizes clearly and distinctly and the deductions we make from them by purely mental processes really apply to the physical world.

As for the study of the physical world, Descartes was sure mathematics would suffice. He says in *The Principles of Philosophy:*

> I frankly confess that in respect to corporeal things I know of no other matter than that . . . which the geometers entitle quantity and take as being the object of their demonstrations. In treating it I consider only the divisions, the shapes and the movements, and in short admit nothing as true save what can be deduced from those common notions (the truth of which cannot be doubted) with the same evidence as is secure in mathematical demonstration. And since in this manner we can explain all the phenomena of nature, I do not think that we should admit any additional physical principles, or that we have the right to look for any other.

Although Descartes glorified mathematical methodology and certainly thought he could reduce all of science to mathematics, surprisingly he had little use for mathematics proper. Apart from results he communicated in letters, he wrote only one brief book on mathematics, the famous "La Géométrie," in which, independently of Fermat, he created analytic geometry, and this was one of three appendices to his great philosophical work, the *Discourse on Method*. In a letter to Marin Mersenne of July 27, 1638, Descartes wrote:

> I have resolved to quit only abstract geometry, that is to say, the consideration of questions that serve only to exercise the mind, and this, in order to study another kind of geometry, which has for its object the explanation of the phenomena of nature.

Descartes says further in his *Discourse* that the knowledge of physics caused him

> to see that it is possible to attain knowledge which is very useful to life, and that, instead of that speculative philosophy which is taught in the schools, we may find a practical philosophy by means of which, knowing the force and action of fire, water, air, the stars, heavens and all other bodies that environ us . . . and thus render ourselves the masters and processors of nature.

One of the appendices to his *Discourse* indicates some interest in applications. In the "Dioptric" he sought to improve the telescope and microscope. He also made studies in biology, and despite his emphasis on the power of the mind, he did some experimenting.

Because Descartes regarded the external world as consisting only of matter in motion, how could he account for tastes, smells, colors, and the qualities of sounds? Here Descartes adopted an older Greek doctrine, the doctrine of primary and secondary qualities. As stated by Democritus (*c.* 460–*c.* 370 B.C.), the doctrine maintained that "Sweet and bitter, cold and warm as well as the colors, all these things exist but in opinion and not in reality; what really exists are unchangeable particles, atoms, and their motions in empty space." The primary qualities, matter and motion, exist in the physical world; the secondary qualities, taste, smell, color, warmth, and the pleasantness or harshness of sounds are only effects that the primary qualities induce in the sense organs of human beings by the impact of external atoms on these organs.

Descartes illustrated this distinction between primary and secondary qualities by using a piece of beeswax. He noted that it has sweetness, odor, color, shape, and size, and also that it is hard and cold. When struck it emits a note. Suppose, however, that one places it near a fire. Its taste and smell vanish, the color changes, the shape is altered, the size increases, and it becomes liquid and hot. It emits no sound when struck. In short, practically all of its properties change, and yet it is the same piece of beeswax. What identifies it as the same object?

The mind, going beyond the senses, recognizes the extension and motion of the beeswax as basic.

Thus to Descartes there are two worlds: one a huge mathematical machine existing in space and the other the world of thinking minds. The effect of elements in the first world on the second produced the nonmathematical or secondary qualities of matter. The real world is the totality of mathematically expressible motions of objects in space and time, and the entire universe is a great, harmonious, and mathematically designed machine.

Even cause and effect were explained by Descartes in purely mathematical terms. The relationship was that of a theorem that has been deduced from previous theorems and axioms. The new theorem—the effect—is no more than a logical consequence of the old in a pattern predetermined by the axioms. *Cause sive ratio;* "cause is nothing but reason." To the senses, cause and effect follow each other in time, and one seems to force the other in some way; however, this appearance in temporal succession and the impression of physical necessity are due to the limitations of the senses.

He had, however, to find the simple, clear, and distinct truths that would play the same part in his philosophy that axioms play in mathematics proper. The results of his search are famous. From the one reliable source that his doubts left unscathed—his consciousness of self—he extracted the building blocks of his philosophy: (a) I think, therefore I am; (b) each phenomenon must have a cause; (c) an effect cannot be greater than the cause; and (d) the ideas of perfection, space, time, and motion are innate in the mind.

Because humans doubt so much and know so little, they are not perfect beings. Yet, according to Descartes's axiom (d) above, the human mind does possess the idea of perfection and, in particular, of an omniscient, omnipotent, eternal, and perfect Being. How do these ideas come about? In view of axiom (c), the idea of a perfect Being could not be derived from or created by the imperfect human mind; hence it could be obtained only from the existence of a perfect Being, who is God. Therefore, God exists.

A perfect God would not deceive us, and so our intuition can be trusted to furnish some truths. Hence the axioms of mathematics, for example, our clearest intuitions, must be truths. The theorems, as already noted, must be correct, because God would not allow us to reason falsely.

The knowledge of nature should be put to use for the good of humanity. To those who asserted that mathematics provides an opportunity to exercise one's inventive faculties and the satisfaction of deploying ingenuity and triumphing over subtle difficulties, Descartes replied that thanks to the new algebraic method (he refers to analytic geometry, which he and Pierre de Fermat independently created), mathematics has become a mechanical science within anyone's reach.

Although Descartes's philosophical and scientific doctrines subverted Aristotelianism and scholasticism, in one fundamental respect he was a scholastic because he drew from his own mind propositions about the nature of being and reality. He believed in a priori truths and that by its own power the intellect may arrive at a perfect knowledge of all things. Thus he stated laws of motion on the basis of a priori reasoning. However, he did promulgate a general and systematic philosophy that shattered the hold of scholasticism and opened up fresh channels of thought. By reducing natural phenomena to purely physical happenings, he did much to rid science of mysticism and occult forces. Moreover, because he tackled nearly all the scientific problems of his day with bold and innovative concepts and methods, he stimulated others to create new scientific theories. Descartes's writings were highly influential in the second half of the seventeenth century. His deductive and systematic philosophy impressed his contemporaries, most especially Newton, with the importance of motion. Daintily bound expositions of Descartes's philosophy adorned ladies' dressing tables.

The supremacy of human reason, the invariability of natural laws, the doctrine of extension and motion as the essence of physical objects, the distinction between body and mind, and the distinction between qualities that are real and inherent in objects and qualities that are only apparently present but are actually due to the reaction of the mind to sense data are elaborated in Descartes's writings and have been influential in shaping modern thought.

It is not our purpose here to elaborate on the philosophical paths that Descartes followed, however worthy of study they are in their own right. What is relevant to our story is that the truths of mathematics and the mathematical method guided a great thinker groping his way through the intellectual upheavals of the seventeenth century. His philosophy may indeed be characterized as mathematized philosophy. It is far less mystical, metaphysical, and theological, and far more rational, than those of his medieval and Renaissance predecessors. By carefully examining the meaning and reasoning involved in all of his mathematical steps, he taught us to look within ourselves for truths and to cast off our pupilage to antiquity and authority. With Descartes, theology and philosophy parted company.

Galileo (1564–1642), whom we have already met as a defender of the heliocentric theory, also offered a philosophy of science that agrees in large part with Descartes's but that proved to be the more radical and more effective procedure. Galileo's grand plan for reading the book of nature proposed a totally new concept of scientific goals and of the role of mathematics in achieving them. It is Galileo's plan for studying and mastering nature that inaugurated modern mathematical physics.

What led Galileo to his revolutionary ideas on the methodology of science is unclear. He knew that Ptolemy had called his own geocentric theory just a mathematical scheme, and he knew that Copernicus had involved mathematical simplicity to defend his heliocentric theory. (Kepler did likewise, but Galileo ignored Kepler's work.) Galileo agreed with both Copernicus and Ptolemy—the world is designed mathematically. In his now famous book, *The Sidereal Messenger* (1610), he writes:

> Philosophy [nature] is written in that great book which ever lies before our eyes. I mean the universe, but we cannot understand it if we do not first learn the language and grasp the symbols in which it is written. The book is written in the mathematical language, and the symbols are triangles, circles and other geometrical figures without whose help it is humanly impossible to comprehend a single word of it, and without which one wanders in vain through a dark labyrinth.

Nature is simple and orderly; its behavior, regular and necessary. It acts in accordance with perfect and immutable mathematical laws. Divine reason is the source of the rational in nature. God put into the world rigorous mathematical necessity that humans, although their reason is related to God's, reach only laboriously. Mathematical knowledge is therefore not only absolute truth but as sacrosanct as any line of the Scriptures. In fact, it is superior for there is much disagreement about the Scriptures, but there can be none about mathematical truth. Moreover, the study of nature is as devout as the study of the Bible: "Nor does God less admirably reveal Himself to us in Nature's actions than in the Scriptures' sacred dictions."

Although Descartes had made a move toward finding laws of motion, he did not face squarely some problems raised by the introduction of the heliocentric theory. Under this theory the Earth was both rotating and revolving around the sun. Why then should objects stay with the Earth? Why should dropped objects fall to Earth if Earth was no longer the center of the universe? Moreover, all motions, projectile motion for example, seemed to take place as though the Earth were at rest. New principles of motion were needed to account for these terrestrial phenomena.

The bold new plan, proposed by Galileo and pursued by his successors, is that of obtaining *quantitative descriptions* of scientific phenomena *independently of any physical explanations.* Galileo's plan may be clarified by an example. In the simple situation in which a ball is dropped from a person's hand one might speculate endlessly on *why* the ball falls. Galileo advised us to do otherwise. The distance the ball falls from its starting point increases as time elapses from the instant it is dropped. In mathematical language, the distance the ball falls and the time that elapses as it falls are called *variables,* because both change as the ball falls. Galileo sought some mathematical relationship between these variables. The answer he obtained is written today in that scientific shorthand known as a formula. For our example, this formula is $d = 16t^2$. This formula states that the number of feet (d) that the ball falls in t seconds is 16 times the square of the number of seconds. For example, in 3 seconds the ball falls 16×3^2 or 144 feet; in 4 seconds, the ball falls 16×4^2 or 256 feet, and so forth.

Notice that the formula is compact, precise, and quantitatively complete. For each value of one variable, *time* in this case, the corresponding value of the other variable, *distance,* may be calculated exactly. This calculation can be performed for millions of values of the time variable, actually an infinite number of values, so that the simple formula $d = 16t^2$ contains an infinite amount of information.

But an important distinction must be made: this mathematical formula is a description of what occurs and not an explanation of a causal relationship. The formula $d = 16t^2$ says nothing about why a ball falls. It merely gives quantitative information about how a ball falls. Furthermore, even though such formulas are used to relate variables that the scientist suspects are causally related, it is nevertheless true that he does not have to investigate, or understand, the causal connection to treat the situation successfully. It is this fact that Galileo saw clearly when he emphasized mathematical description against the less successful qualitative and causal inquiries into nature.

It was Galileo's decision, then, to seek the mathematical formulas that describe nature's behavior. This thought, like most thoughts of genius, may leave the reader initially unimpressed. There seems to be no real value in these bare mathematical formulas. They explain nothing. They simply *describe* in precise language. Yet such formulas have proved to be the most valuable knowledge human beings have ever acquired about nature. We shall see that the amazing practical as well as theoretical accomplishments of modern science have been achieved mainly through the quantitative, descriptive knowledge that has been amassed and manipulated rather than through metaphysical, theological, and even mechanical explanations of the causes of phenomena.

Galileo says in his *Discourses and Mathematical Demonstrations Concerning Two New Sciences* (1638): "The cause of the acceleration of the motion of falling bodies is not a necessary part of the investigation." More generally, he points out that he will investigate and demonstrate some of the properties of motion without regard to what the causes might be. Positive scientific inquiries must be separated from questions of ultimate causation, and speculation as to physical causes must be abandoned. Galileo might well have admonished scientists: Yours not to reason why; yours but to quantify.

First reactions to this thought of Galileo are likely to be negative. Descriptions of phenomena in terms of formulas hardly seem to be more than a first step. It would seem that the true function of science had really been grasped by the Aristotelians, namely, that we should attempt to explain physically why phenomena happened. Even Descartes protested Galileo's decision to seek descriptive formulas: "Everything that Galileo says about bodies falling in empty space is built without foundation; he ought first to have determined the nature of weight." Furthermore, said Descartes, Galileo should reflect about ultimate reasons. Yet we now know, in the light of subsequent developments, that Galileo's decision to aim for description was the deepest thought and the most fruitful that anyone has had about scientific methodology. Its significance, which will be more fully apparent later, is that it placed science far more squarely into the domain of mathematics.

The decision to seek the formulas that describe phenomena leads in turn to the question: What quantities should be related by formulas? A formula relates the numerical values of varying physical entities; hence these entities must be measurable. The principle Galileo followed next was to measure what is measurable and to render measurable what is not yet so. His problem then became that of isolating those aspects of natural phenomena that are basic and capable of measurement.

Descartes had already fixed on matter moving in space and time as the fundamental phenomenon of nature. Galileo therefore sought to isolate the characteristics of matter in motion that could be measured and then be related by mathematical laws. By analyzing and reflecting on natural phenomena, he decided to concentrate on such concepts as space, time, weight, velocity, acceleration, inertia, force, and momentum. In the selection of these particular properties and concepts Galileo again showed genius, for the ones he chose are not immediately discernible as the most important nor are they readily measurable. Some, such as inertia, are not even obviously possessed by matter; their existence had to be inferred from observations. Others, such as momentum, had to be created. Yet these concepts did prove to be most significant in unearthing many of the secrets of nature.

Another element in the Galilean approach to science proved equally important in the sequel. Science was to be patterned on the mathematical model. Galileo and his immediate successors felt sure that they could find some laws of the physical world that would appear to be as unquestionably true as the axiom of Euclid that a straight line may be drawn through any two points. Perhaps contemplation, experimentation, or observation would suggest these axioms of physics; at any rate, once they were discovered their truth would be intuitively evident. With such fundamental intuitions Galileo, in this respect like Descartes, hoped to deduce a number of other truths in precisely the manner in which Euclid's theorems followed from his axioms.

However, Galileo departed radically from the Greeks, the medievalists, and even Descartes in the method of obtaining first principles. The pre-Galileans and Descartes had believed that the mind supplies the basic principles. The mind had but to think about any class of phenomena and it would immediately recognize fundamental truths, as was clearly evidenced in mathematics. Axioms such as *equals added to equals give equals* and *two points determine a line* suggested themselves immediately in thinking about number or geometrical figures and were indubitable truths. So, too, did the Greeks find some physical principles equally appealing. That all objects in the universe should have a natural place was no more than fitting. The state of rest seemed clearly more natural than the state of motion. Because the heavenly bodies were perfect and repeated their motions in definite periods of time, and because the circle was the perfect curve that allowed repeated behavior, it was clear that the heavenly bodies must move in circles or, at worst, combinations of circles. It seemed indubitable, too, that to put and keep bodies in motion force must be applied. To believe that the mind supplies fundamental principles does not deny that observations might enter into the process of obtaining these principles. Yet the observations merely evoke the correct principles, just as the sight of a familiar face may call to mind facts about that person.

Galileo stressed that the way to obtain correct and basic principles is to pay attention to what nature says rather than what the mind prefers. He openly criticized scientists and philosophers who accepted laws that conformed to their preconceived ideas as to how nature must behave: Nature did not first make human brains and then arrange the world so that it would be acceptable to human intellects. To the medievalists who kept repeating Aristotle and debating what he meant, Galileo addressed the criticism that knowledge comes from observation and not from books. It was useless to debate the words of Aristotle. Those who did he called paper scientists who fancied that science was to be studied like the *Aeneid* or the *Odyssey* or by the collation of texts. Nature made things as she liked; let human reason struggle as best it could to understand her ways. "Nature nothing careth whether her abstruse reasons and methods of operating be or be not exposed to the capacity of men. . . . When we have the decrees of nature authority goes for nothing."

Such criticisms had been voiced by many predecessors of Galileo. Leonardo da Vinci had said that sciences that arise and end in thought do not give truths because in these mental considerations no experience enters, and with this nothing is sure. "If you do not rest on the good foundation of nature you will labor with little honor and less profit." Galileo's contemporary, Francis Bacon, had spoken sharply in favor of banishing the various idols that preoccupied the human mind and prevented people from seeing the truth. However, before Galileo the use of experience to obtain basic principles was fumbling and without direction.

Yet the modernist Descartes did not grant the wisdom of Galileo's reliance on experimentation. The facts of the senses, Descartes said, can only lead to delusion. Reason penetrates such delusions. From the innate general principles supplied by the mind we can deduce particular phenomena of nature and understand them. In much of his scientific work Descartes did experiment and require that theory fit facts, but in his philosophy he was still tied to truths of the mind.

Even though Galileo experimented purposefully and with telling effect, we must not conclude that experimentation was undertaken on a large scale and that it became a new and decisive force in science. This was not to come until the nineteenth century. There were, of course, some famous experimenters in the seventeenth century: Robert Hooke, the physicist; Robert Boyle, the chemist; and Christian

Huygens, the mathematician and physicist, to say nothing of Galileo himself and Sir Isaac Newton. Galileo, a transitional figure insofar as experimentation is concerned, was not the experimenter he is often reputed to be. He, and even Newton, believed that a few critical experiments and keen observation would readily yield correct fundamental principles. Newton emphasized his reliance on mathematics and said he used experiments largely to make his *results* physically intelligible and to convince the "vulgar." Many of Galileo's so-called experiments were really "thought-experiments"; that is, he used his experience to imagine what would happen if an experiment were performed and then drew an inference as confidently as if he had actually carried out the experiment. In his writing he describes experiments he never made. He advocated the heliocentric theory even though, in the stage in which Copernicus left it, the theory did not yet accord well with observations. In describing some experiments on motion along an inclined plane, Galileo did not give actual data but said that the results agreed with theory to a degree of accuracy that is incredible in view of the poor clocks available in his time. A few fundamental principles derived from nature and much mathematical reasoning constituted Galileo's method. When in his *Dialogue on the Great World Systems* he describes the motion of a ball dropped from the mast of a moving ship, he is asked by Simplicio, one of the characters, whether he had made an experiment. Galileo replies, "No, and I do not need it, as without any experience I can confirm that it is so because it cannot be otherwise." He says in fact that he experimented rarely and then primarily to refute those who did not follow the mathematical method.

Galileo did have some preconceptions about nature that made him confident a few experiments would suffice. For example, when he undertook to study accelerated motion—that is, motion with changing velocity—he assumed as the simplest principle that the increases in velocity in equal intervals of time were equal. This he called uniformly accelerated motion. Thus for Galileo the deductive mathematical part of the scientific enterprise played a greater part than the experimental. He was prouder of the abundance of theorems that flow from a single principle than of the discovery of the principle itself. And so we see a pattern: the scientists who fashioned modern science—and we can include Descartes, Galileo, Huygens, Newton, and also Copernicus and Kepler here—approached the study of nature as mathematicians in their general method and in their concrete investigations. Primarily speculative thinkers, they expected to apprehend broad, deep, but also simple, clear, and immutable mathematical principles, either through intuition or through crucial observations and experiments, and then expected to deduce new laws from these fundamental truths entirely in the manner in which mathematics proper had constructed its geometry. Deductive reasoning made up the bulk of the activity, and whole systems of thought were to be so derived.

The expectation of Galileo that just a few experiments would suffice can be readily appreciated. Because these men were convinced that nature is mathematically designed, they saw no reason for not proceeding in scientific matters as mathematicians had proceeded in their domain. As John Herman Randall says in his *The Making of the Modern Mind,*

> Science was born of a faith in the mathematical interpretation of nature.... Modern science arose as, and in fact was known as, natural philosophy, the inclusion of the word philosophy being more than an accident and actually descriptive of the approach used. It is the approach of thinkers who rely essentially on reason and, in the present case, on mathematical principles and procedures as the primary tool of reason.

Nevertheless, Galileo's doctrine that physical principles must rest on experience and experiments was revolutionary and crucial. Galileo himself had no doubt that the true principles—those used by God to fashion the universe—could still be obtained, but by opening the door to the role of experience he allowed the devil of doubt to slip in unnoticed. For if the basic principles of science must come from experience, why not the axioms of mathematics? This question did not trouble Galileo or his successors until 1800. Mathematics would enjoy a privileged position until that time.

To get to the heart of phenomena, Galileo urged and practiced one more principle, namely, idealization. By this he meant that one should ignore trivial and minor factors. Thus, a ball falling to earth encounters air resistance, but for a fall of, say, a few hundred feet the air resistance is slight and can for almost all purposes be neglected. Similarly, a somewhat compact object has size and shape, but essentially it can be treated as a point mass; that is, the mass is regarded as though it were all concentrated at one point. He also ignored secondary qualities such as taste, color, and smell as opposed to size, shape, quantity, and motion. In other words, he adopted the philosophical doctrine that distinguished between primary and secondary qualities of matter. He said:

> White or red, bitter or sweet, sounding or silent, sweet-smelling or evil-smelling are names for certain effects upon the sense-organs; they can no more be ascribed to the external objects than can be tickling or the pain

> caused sometimes by touching these objects. . . . If ears, tongues, and
> noses are removed, I am of the opinion that shape, quantity, and motion
> would remain but there would be an end of smells, tastes, and sounds,
> which abstracted from the living creature, I take to be mere words.

Shape, quantity (size), and motion, then, are the primary or physically
basic properties of matter. These are real and external to human
perception.

Thus what Galileo advocated was the stripping away of incidental
or minor effects to get at the major one. He started from observations
and then imagined what would happen if all resistance were
removed—that is, if bodies *fell in a vacuum*—and he obtained the
principle that in a vacuum all bodies fall according to the same law.
Having observed that the motion of a pendulum is little affected by air
resistance, he then experimented with pendulums to confirm his principles. Likewise, having suspected that friction too is a secondary
effect, he experimented with smooth balls rolling down a smooth slope
to obtain laws about frictionless motion. So Galileo did not just experiment and infer from experiments. He discarded from his interpretations of the experiments what was relatively unimportant. His great-
ness consisted, in part, in his asking the proper questions about nature.

Of course, actual bodies do fall in resisting media. What could
Galileo say about such motions? His answer was:

> [H]ence, in order to handle this matter in a scientific way, it is necessary
> to cut loose from these difficulties [air resistance, friction, etc.] and having
> discovered and demonstrated the theorems, in the case of no resistance,
> to use them and apply them with such limitations as experience will
> teach.

By stripping away air resistance and friction and seeking basic
laws for motion in a vacuum, Galileo not only contradicted Aristotle
and even Descartes by thinking of bodies moving in empty space, he
also used the method of idealizing or abstracting the essential properties for his purposes. He did just what the mathematician does in
studying real figures. The mathematician strips away molecular structure, color, and thickness of lines to get at some basic properties, and
then concentrates on these. Similarly, Galileo penetrated to basic physical factors. The mathematical method of abstraction is indeed a step
away from reality, but paradoxically, it leads back to reality with
greater power than if all the factors actually present are taken into
account at once.

Galileo showed wisdom in still another tactic. He did not try, as did scientists and philosophers before him, to embrace all natural phenomena. He selected a few basic phenomena and studied these intensively. He deemed it wise to go cautiously and circumspectly. Galileo showed the restraint of the master.

The Galilean plan contained, then, four main features. The first was to seek quantitative descriptions of physical phenomena and embody these in mathematical formulas. The second was to isolate and measure the most fundamental properties of phenomena. These would be the variables in the formulas. The third was to build up science deductively on the basis of fundamental physical principles. The fourth was to idealize.

To put this plan into execution, Galileo had to find fundamental laws. One might obtain a mathematical formula relating the number of marriages in Siam and the price of horseshoes in New York City, as these quantities vary from year to year. Such a formula is of no value to science, however, for it does not encompass, either directly or by implication, any useful information. The search for fundamental laws was another immense task, because once again Galileo had to break with his predecessors. His approach to the study of matter in motion had to take into account an Earth moving through space and rotating on its axis, and these facts in themselves invalidated much of the only significant system of mechanics that the Renaissance world possessed, namely, the mechanics of Aristotle.

Galileo was at first inclined to accept the Aristotelian law that heavier bodies fall to earth faster than light ones. Then it occurred to him to ask: Suppose I break the heavy one into two pieces. Would they now fall as two lighter bodies? Suppose further that I tie them together or glue them together. Then what? Are they now two pieces or one? He concluded that all bodies fall at the same speed if air resistance is neglected.

According to Aristotle, a force is required to keep a body in motion. Hence, to keep an automobile or a ball moving, even on a very smooth surface, some propelling force should be present. Galileo had greater insight into this phenomenon than Aristotle. Actually a rolling ball or moving automobile is hindered somewhat by the resistance of air and retarded by friction between it and the surface on which it rolls. If these hindering actions were not present, *no* propelling force would be needed to keep the automobile rolling. It would continue at the same speed *indefinitely;* moreover, it would follow a straight-line path. This fundamental law of motion—that a *body undisturbed by forces will continue indefinitely at a constant speed and in a straight line*—was independently discovered by Galileo (and also stated by Descartes) and is now known as Newton's first law of motion. The law says that a body will change its speed only if it is acted on by a force. Thus, bodies possess the property of resisting change in speed. This property of matter, namely, resistance to change in speed, is called its *inertial mass* or simply its mass.

This very first principle contradicts the related one of Aristotle. Does this mean that Aristotle made obvious blunders, or that his observations were too crude or too few to yield the correct principle? Not at all. Aristotle was a realist, and he taught what observations actually do suggest. Galileo's method, however, was more sophisticated and consequently more successful. Galileo approached the problem as a mathematician. He idealized the phenomenon by ignoring some facts to favor others just as the mathematician idealizes the stretched string and the edge of a ruler by concentrating on some properties to the exclusion of others. By ignoring friction and air resistance and by imagining motion to take place in a pure Euclidean vacuum, he discovered the correct fundamental principle.

What can be said about the motion of a body if some force is applied to it? Here Galileo made a second fundamental discovery: the continuous application of a force causes a body to gain or lose velocity. Let us call the gain or loss in velocity per unit of time the acceleration of the body. Thus, if a body gains or loses velocity at the rate of 30 feet per second each second, its acceleration is 30 feet per second each second, or in abbreviated form, 30 ft/sec^2.

For example, a constant air resistance causes a constant loss in velocity, and this accounts for the fact that an object rolling or sliding on a smooth floor will lose velocity continually until it has zero velocity. Conversely, if a moving object does possess acceleration, then some force must be acting. An object falling to the earth from some height does possess acceleration. In Galileo's time the notion had already gained some acceptance that this force must be the pull of the Earth, but without wasting much time on speculation about this notion, Galileo investigated the quantitative facts about falling bodies.

He discovered that if air resistance is neglected, all bodies falling to the surface of the Earth have the same constant acceleration *a*; that is, they gain velocity at the same rate, 32 feet per second each second. In symbols,

$$a = 32 \tag{1}$$

If the body is dropped—that is, merely allowed to fall from the hand—it will start with zero velocity. Hence, at the end of 1 second its velocity is 32 feet per second; at the end of 2 seconds its velocity is 32 times 2 or 64 feet per second; and so forth. At the end of t seconds its velocity v is $32t$ feet per second; in symbols,

$$v = 32t \qquad (2)$$

This formula tells us exactly how the velocity of a falling body increases with time. It says, too, that a body that falls for a longer time will have a greater velocity. This is a familiar fact, for most people have observed that bodies dropped from high altitudes hit the ground at higher speeds than bodies dropped from low altitudes.

We cannot multiply the velocity by the time to find the distance that a dropped body falls in a given amount of time. This would give the correct distance only if the velocity were constant. Galileo proved, however, that the correct formula for the distance d the body falls in t seconds is

$$d = 16t^2 \qquad (3)$$

where d is the number of feet the body falls in t seconds. For example, in 3 seconds, the body falls 16×3^2 or 144 feet.

By dividing both sides of formula (3) by 16 and then taking the square root of both sides, we find that the time required for an object to fall a given distance d is given by the formula $t = \sqrt{d/16}$. Notice that the mass of the falling body does not appear in this formula. We can thus see that all bodies take the same time to fall a given distance. This is the lesson Galileo is supposed to have learned by dropping objects from the tower of Pisa. People still find it difficult to believe, nevertheless, that a piece of lead and a feather when dropped from a height in a vacuum reach the ground in the same time.

The acceleration of thirty-two feet per second each second, which all objects falling to earth possess, is caused by the force of gravity. When we speak of this force as applied to objects near the surface of the Earth, we call it weight. Although Galileo did not relate weight and mass, we should note that the weight (w) of any object on the Earth is always thirty-two times its mass (m). In symbols,

$$w = 32m \qquad (4)$$

Thus, two distinct properties of objects, weight and mass, are simply related in that one is always thirty-two times the other. Because of this constant relationship we tend to confuse the two properties, but we should be clear about the distinction. Mass is the property of resistance to change in speed or direction; weight is the force with which the Earth attracts the object. If an object rests on a horizontal surface, the surface opposes the force of gravity. Hence, insofar as motion along the surface is concerned, the weight of the object plays no role. Nevertheless the mass of the object is still present. We shall see in a later chapter how important it is to distinguish between mass and weight.

We owe to Descartes, the profound and highly influential philosopher, the emphasis on mathematics that placed it at the forefront of the sciences from the seventeenth century onward, but it was Galileo's methodology that enabled humanity to uncover the behavior of many natural phenomena that might otherwise have remained unknown.

We could pursue many more of Galileo's specific mathematical achievements, such as his mathematical description of projectile motion, but our chief concern with his work is his methodology. With his *Discourses* of 1638 Galileo launched modern physical science on its mathematical course, founded the modern science of mechanics, and set the pattern for all of modern scientific thought. As we shall soon see, Newton took over Galileo's methodology and gave unsurpassed demonstrations of its effectiveness.

Mathematics and the Mystery of Gravitation

> *I have not yet been able to discover the cause of these properties of gravity from phenomena and I frame no hypotheses. . . . It is enough that gravity does really exist and acts according to the laws I have explained, and that it abundantly serves to account for all the motions of celestial bodies.*   Isaac Newton

In 1642, the very year of Galileo's death, on a farm located in a secluded English hamlet, a woman recently widowed gave birth to a frail and premature child. From such an insignificant origin and with a body so weak that his life was despaired of, Isaac Newton lived to be eighty-five and to acquire fame as great as any man's. As we shall see, Newton, using essentially Galileo's methodology, would take over where Galileo left off. As Alfred North Whitehead once put it, "Galileo represents the assault and Newton the victory."

Except for a strong interest in mechanical contrivances, Newton showed no special promise as a youth. For the negative reason that he showed no interest in farming, his mother sent him to Cambridge, and he entered Trinity College in 1661. Despite several advantages of attendance there—such as the opportunity to study the works of Descartes, Copernicus, Kepler, and Galileo, and the opportunity to listen to the famous mathematician Isaac Barrow—Newton seemed to profit little. He was even found to be weak in geometry and at one time almost changed his course of study from natural philosophy to law. Four years of undergraduate study ended as unimpressively as they began.

At that point an outbreak of the plague in the area around London led to the closing of the university. Newton therefore spent the years

1665 and 1666 in the quiet of the family home at Woolsthorpe. During this period he initiated his great work in mechanics, mathematics, and optics. He realized that the law of gravitation, which we shall examine shortly, was the key to an embracing science of mechanics; he obtained a general method for treating the problems of the calculus; and through experiments he made the epochal discovery that white light such as sunlight is really composed of all colors from violet to red. "All this," Newton said later in life, "was in the two plague years of 1665 and 1666, for in those days I was in the prime of my age for invention, and minded mathematics and philosophy [science] more than at any other time since."

Newton returned to Cambridge in 1667 and was elected a Fellow of Trinity College. In 1669 Isaac Barrow resigned his professorship of mathematics to devote himself to theology, and Newton was appointed in Barrow's place. Newton apparently was not a successful teacher, for few students attended his lectures; nor did anyone comment on the originality of the material he presented.

In 1684 his friend Edmond Halley, the astronomer of Halley's comet fame, urged him to publish his work on gravitation and even assisted him editorially and financially. Thus, in 1687, the classic of science, *The Mathematical Principles of Natural Philosophy,* often briefly referred to as the *Principia* or the *Principles,* was published.

After the publication Newton did receive widespread acclaim. The *Principles* went through three editions, and popularizations became common. Actually the *Principles* needed popularization, for the book is extremely difficult to read and is not at all clear to laypeople, despite statements by educators to the contrary. The greatest mathematicians worked for a century to elucidate fully the material of the book.

Newton gave due credit to his predecessors, and he did not believe that his work was of incomparable importance. When quite old he told his nephew:

> I do not know what I may appear to the world; but to myself I seem to have been only like a boy playing on the seashore, and diverting myself in now and then finding a smoother pebble or a prettier shell than ordinary, whilst the great ocean of truth lay all undiscovered before me.

Of the great contributions of his youth, Newton's philosophy of science and his work on gravitation are most relevant to our present subject. The philosophy stated more explicitly the program for science that Galileo had initiated: from clearly verifiable phenomena, laws are to be framed that state nature's behavior in the precise language of mathematics. By the application of mathematical reasoning to these laws, new ones are to be deduced. Like Galileo, Newton wished to know how the Almighty had fashioned the universe, but he did not hope to fathom the mechanism behind many phenomena.

In the preface to his *Principia* he says:

> Since the ancients (as we are told by Pappus) esteemed the science of mechanics of greatest importance in the investigation of natural things, and the moderns, rejecting substantial forms and occult qualities, have endeavored to subject the phenomena of nature to the laws of mathematics, I have in this treatise cultivated mathematics as far as it relates to philosophy [science], and therefore I offer this work as the mathematical principles of philosophy, for the whole burden in philosophy seems to consist in this—from the phenomena of motions to investigate the forces of nature and then from these forces to demonstrate the other phenomena; and to this end the general propositions in the first and second Books are directed.

Of course, mathematical principles to Newton as to Galileo were quantitative principles. As Newton says in his *Principia,* his purpose is to discover and put forth the exact manner in which "all things had been ordered in measure, number and weight."

In this task of describing nature, Newton's most famous contribution was to unite heaven and Earth. Galileo had viewed the heavens as no one had previously been able to, but his successes in describing nature mathematically were limited to motions on or near the surface of the Earth. During Galileo's lifetime his contemporary, Kepler, had obtained his three famous mathematical laws on the motions of the heavenly bodies and had thereby simplified the heliocentric theory. The two branches of science, terrestrial and celestial, seemed to be independent of each other. The challenge to find some relationship between them stirred the great scientists. It was met by the greatest one.

There was good reason to believe that some unifying principle did exist. Under Galileo's first law of motion, which Newton may have learned from Descartes's writings, or from Galileo whom Newton credits, bodies should continue to move in straight lines unless disturbed by forces. Hence the planets, set into motion somehow, should move in straight lines whereas, according to Kepler, they moved in ellipses around the sun. Some force must therefore be deflecting the planets continually from straight-line paths, just as a weight swung at the end of a string does not fly off in a straight line because the hand exerts a force pulling it in. Presumably the sun itself was acting as an attracting force on the planets. The scientists of Newton's day also appreciated that the Earth attracts bodies to it. Since both the Earth and the sun attract bodies, the idea of unifying both actions under one theory was advanced and discussed even in Descartes's day.

Newton converted a common thought into a mathematical problem and, without determining the physical nature of the forces involved, solved this problem by brilliant mathematics. Story has it that the identity of the Earth's pull on objects and the sun's pull on the Earth was brought to Newton's attention by the fall of an apple from a tree. Karl Friedrich Gauss, the mathematician, said that Newton told this story to dispose of stupid persons who asked him how he discovered the law of gravitation, but the story is authentic. At any rate, this apple, unlike another that played a role in history, improved the status of humanity.

Newton began by considering the problem of sending a projectile out horizontally from the top of a mountain. Of course, Galileo had worked out just such problems and had shown that the resulting path would be a parabola. Moreover, if the horizontal velocity of the projectile were greater, the path would still be a parabola but a wider one because the projectile would travel farther out horizontally. But Galileo had considered projectiles that did not travel far, and he had therefore ignored the curvature of the Earth. Hence Newton's first thought was that if a projectile were shot out horizontally with enough speed, it might follow the curved path VD (Figure 26). Would it not move off into space and escape from the Earth altogether? No, because the Earth would continue to attract it. In what direction would the Earth pull the

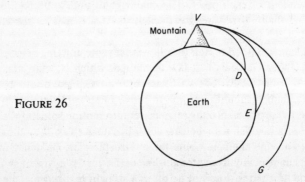

FIGURE 26

projectile? Galileo had always considered that gravity pulls all objects straight down, but straight down for an object that is circling the Earth means toward the center of the Earth. Hence, projectiles shot out from the mountaintop would be pulled in toward the Earth. If shot out with greater velocity, a projectile might take the path *VE* and if shot out with a sufficiently large velocity, might travel around the Earth and perhaps continue to circle it indefinitely. And so Newton argues in his *Mathematical Principles:*

> And after the same manner that a projectile, by the force of gravity, may be made to revolve in an orbit, and go around the whole earth, the moon also, either by the force of gravity, if it is endowed with gravity, or by any other force, that impels it toward the earth, may be continually drawn aside towards the earth, out of the rectilinear way which by its innate force [inertia] it would pursue; and would be made to revolve in the orbit which it now describes.

If the Earth by its gravitational attraction could cause the moon to "circle" the Earth, then so might the sun by its gravitational attraction cause the planets to "circle" the sun. Hence Newton had some basis for entertaining the exciting prospect that the same force that pulled objects near the Earth to it also kept the moon moving around the Earth and the planets moving around the sun.

All of Newton's reasoning thus far was qualitative and speculative. To make progress one now had to become quantitative. Referring still to the force acting on the moon, Newton continues:

> If this force were too small, it would not sufficiently turn the moon out of a rectilinear [straight-line] course; if it were too great, it would turn it too much, and draw the moon from its orbit toward the earth. It is necessary that the force be of a just quantity, and it belongs to the mathematicians to find the force that may serve exactly to retain a body in a given orbit with a given velocity.

Newton's reasoning in showing that the same formula applies to earthly and heavenly bodies is now classic. We shall consider a somewhat crude version of it that may nevertheless give the essence. The path of the moon around the Earth is roughly a circle. Because the moon, *M* in Figure 27, does not follow a straight-line path such as *MP,* it evidently is pulled toward the Earth by some force. If *MP* were the distance the moon might have moved in 1 second with no gravitational force acting, then *PM'* is the distance the moon is pulled toward the Earth during that second. Newton used *PM'* as a measure of the Earth's attractive force on the moon. The corresponding quantity in

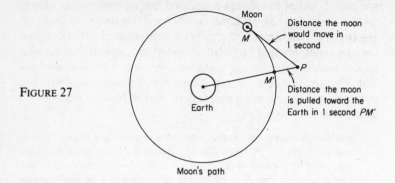

FIGURE 27

the case of a body near the surface of the Earth is 16 feet, for a body dropped from the hand is pulled 16 feet toward the Earth in the first second. Newton wished to show that the same force accounted for both PM' and the 16 feet.

Rough calculation had led him to believe that the force attracting one body to another depends on the square of the distance between the centers of the two bodies involved, and that this force decreases as the distance increases. The distance between the center of the moon and the center of the Earth is about 60 times the radius of the Earth. Hence the effect of the Earth on the moon should be $1/(60)^2$ of its effect on a body near the surface of the Earth; that is, the moon should be pulled toward the Earth $1/(60)^2$ of 16 or 0.0044 feet each second. By using some numerical results obtained by means of trigonometry, Newton found that the moon is pulled toward the Earth by "pretty nearly" that amount each second. Thus he had obtained a most important piece of evidence to the effect that *all* bodies in the universe attract each other in accordance with the same law.

More extensive investigation showed Newton that the precise formula for the force of attraction between *any* two bodies is given by the formula

$$F = \frac{kMm}{r^2} \tag{1}$$

where F is the force of attraction, M and m are the masses of the two bodies, r is the distance between them, and k is the same for all bodies. For example, M could be the mass of the Earth and m, the mass of an object near or on the surface of the Earth. In this case r is the distance from the *center* of the Earth to the object. Formula (1) is, of course, the law of gravitation.

To organize his entire work on terrestrial and celestial motions, Newton stated in his *Principles* what are now referred to as Newton's laws of motions (although the first two were already stated by Descartes and Galileo). The first law states:

> A body remains in a state of rest or constant velocity (speed and direction) when no external force acts on it.

His second law states:

> The net force on a body is the product of its mass and the acceleration produced:

$$F = ma$$

> The acceleration is the increase or decrease in the speed of the body and the change in direction. (In mathematical terms, F and a are vectors.)

His third law states:

> Whenever two bodies interact, the force on the first body due to the second is equal and opposite to the force on the second due to the first.

To these three laws, he added the all-important law of gravitation, formula (1) above. As applied to the planets, it was suggested to Newton by Robert Hooke. However, Newton generalized it to produce a universal law applying to all matter in the universe.

Having obtained some confirmation of the law of gravitation, Newton showed next that the law could be applied to motions on or near the Earth. Here Galileo's work helped him. In mathematical terms, let M be the mass of the Earth and m the mass of an object near the surface. If we rewrite formula (1) as

$$F = \frac{kM}{r^2} m$$

and divide both sides of this equation by m, we obtain

$$\frac{F}{m} = \frac{kM}{r^2} \tag{2}$$

Now, regardless of what object near the Earth's surface we consider, the quantities on the right side of formula (2) are the same, because r is about 4000 miles, M is the mass of the Earth, and k is the same for all bodies.

However, the second law of motion says that any force acting on a body of mass m gives the body an acceleration. In particular, the force of gravity exerted by the Earth on a body should give it an acceleration. The relation of any force to the acceleration it causes is $F = ma$, or

$$\frac{F}{m} = a \tag{3}$$

Hence, when the force F in formula (3) is that of gravity, we can equate the right sides of formulas (2) and (3) because the left sides are then equal; that is,

$$a = \frac{kM}{r^2}$$

This result says that the acceleration imparted to an object by the force of gravity of the Earth is always kM/r^2. Because k is a constant, M is the mass of the Earth, and r is the distance of an object from the center of the Earth, the quantity kM/r^2 is the same for all bodies near the surface of the Earth. Hence all such bodies fall with the same acceleration. This, of course, is the result Galileo had already obtained by inference from experiments, and from this result he proved mathematically that all bodies falling from the same height reach the ground in the same amount of time. Incidentally, the value of a is easily measured and is 32 feet per second each second.

It is of some interest, though perhaps incidental to our main theme, that Newton's third law of motion states that to each force there is an equal and opposite force. Hence if the sun exerts a force on the Earth to keep the Earth in its orbit, the Earth should exert an equal and opposite force on the sun. Yet the sun seems stationary. The answer is a bit of Newtonian mathematics. If we now let m and M be the masses of the Earth and the sun, the force of attraction between them is still

$$F = \frac{kMm}{r^2}$$

And the force the Earth imparts to the sun is

$$F = Ma$$

From these two formulas we have

$$\frac{F}{M} = \frac{km}{r^2}$$

and

$$\frac{F}{M} = a$$

Hence the acceleration the Earth imparts to any object is

$$a = \frac{km}{r^2}$$

where m is the mass of the Earth and r is the distance from the Earth to the object. Because this mass is very much smaller than the mass of the sun, the acceleration the Earth imparts to the sun is minor compared to what the sun imparts to the Earth. The sun does move under the attraction of the Earth and other planets, but its motion is so small that it is ignored. As a corollary to this mathematics, we should note that as the Earth pulls on us, so do we pull on the Earth. But we fall to Earth, whereas the Earth's motion toward us is negligible.

Thus far Newton's contributions to the theory of gravitation may be summarized as follows. By studying the motion of the moon, he had inferred the correct form of the law of gravitation. He then showed that this law and the first two laws of motion sufficed to establish valuable knowledge about the motions of objects on the Earth. He had therefore achieved one of the major goals in Galileo's program, because he had shown that the laws of motion and gravitation were fundamental. Like the axioms of Euclid, they served as the logical basis for other valuable laws. What a triumph indeed it would be to deduce, in addition, the laws of motion for the heavenly bodies.

This triumph was also reserved for Newton. A truly momentous series of deductions made by him showed that all three of Kepler's laws follow from the two basic laws of motion and the law of gravitation.

There is an important corollary to these laws that should be informative to the reader who seeks an explanation of the power of mathematics. The major value of the Newtonian laws lies, as we have just seen, in the fact that they apply to so many varied situations in the heavens and on the Earth. The same quantitative relationships epitomize characteristics common to all. Hence, knowledge of the formulas really represents knowledge about all the situations encompassed by the formulas.

The work of Galileo and Newton was not the end but the beginning of a program for science. Newton himself formulated the program in the preface to his *Mathematical Principles of Natural Philosophy,* the classic that contains the work of his brilliant youth:

> We offer this work as mathematical principles of philosophy [science]; for all the difficulty in philosophy seems to consist in this—from the phenomena of motions to investigate the forces of nature, and then from these forces to demonstrate the other phenomena; . . . for by propositions mathematically demonstrated in the first book, we there derive from the celestial phenomena the forces of gravity with which bodies tend to the sun and several planets. Then, from these forces, by other propositions which are also mathematical, we deduce the motions of the planets, the comets, the moon and the sea. I wish we could derive the rest of the phenomena of nature by the same kind of reasoning from mechanical principles; for I am induced by many reasons to suspect that they may all depend upon certain forces by which the particles of bodies, by some cause hitherto unknown, are either mutually impelled towards each other, and cohere in regular figures, or are repelled and recede from each other.

Like a rock rolling down a steep hill, Newton proceeded to secure fundamental mathematical laws and to deduce their consequences. By procedures similar to those discussed in this chapter, the mass of the sun and the mass of any planet with observable satellites were calculated. The idea of centrifugal force was applied to the motion of the Earth and produced the magnitude of the equatorial bulge of the Earth as well as the consequent variation of the weight of an object from point to point on the Earth's surface. From the knowledge of the observed departure from sphericity of the several planets, it became possible to calculate their periods of rotation. The Earth's tides were shown to be caused by the gravitational attractions of the sun and moon.

Yet a number of irregularities in the motions of the heavenly bodies had been observed and were unaccounted for. For example, although the moon always presents the same face to the Earth, more or less of the region near the edges becomes periodically visible. In addition, increased observational accuracy had revealed that the length of the mean lunar month decreased by about one-thirtieth of a second per century. (Such was the order of accuracy that observation and theory had come to handle.) Finally, small changes in the eccentricities of the planetary orbits had also been observed.

Newton was well aware of many of these irregularities, and in his own studies he tackled the motion of the moon. In Newton's time the position of the moon, as observed from a ship, could be used to determine the ship's longitude. (Seaworthy clocks were not available.) Newton did have this practical application in mind. The moon follows an elliptical path, somewhat as a drunken man follows a straight line: it hurries and lingers, and reels from side to side. Newton was convinced that some of this extraordinary behavior was due to the moon being attracted by the sun as well as by the Earth, causing it to depart from a truly elliptical path. In his *Principles,* Newton did show that some of the irregularities were consequences of the laws of motion and gravitation.

Newton also argued that comets should follow elliptical paths and urged Edmund Halley (1656–1742) to look for comets. Halley gathered data on comets seen in the past. He noticed that apparently the same comet appeared in 1531, 1607, and 1682. By using Newton's theory he predicted that this comet would reappear near the end of 1758 or the beginning of 1759. It was seen on Christmas Day 1758 and passed closest to the sun on March 13, 1759. It was seen last in 1910 and should be visible in 1986. (It has already been sighted by telescope, though still far away.) Its period does vary somewhat, because the planets disturb the comet's path.

However, Newton had no proof that all the observed irregularities in the motions of the moon and the planets resulted from gravitational pulls; so he could not show that the cumulative effect would not ultimately disrupt the solar system. Newton's eighteenth-century successors undertook to study these irregularities.

The path of each planet around the sun, as Newton knew, would be an ellipse only if the one planet and the sun were in the heavens. However, the solar system contains nine planets, many with moons, all not only moving around the sun but attracting each other in accordance with Newton's universal law of gravitation. Their motions, therefore, certainly could not be truly elliptical. Their exact paths would be known if it were possible to solve the general problem of

determining the motion of an arbitrary number of bodies, each attracting all the others under the action of gravitation—but this problem is
beyond the capacity of any mathematician. Two of the greatest mathematicians of the eighteenth century did, however, make phenomenal
steps along these lines.

The Italian-born Joseph-Louis Lagrange (1736–1813), in a brilliant exhibition of youthful genius, tackled the mathematical problem
of the moon's motion under the attraction of the sun and Earth, and
solved it at the age of twenty-eight. He showed that the variation in
the portion of the moon that is visible is caused by the equatorial
bulges of both the Earth and the moon. In addition, the attraction of
the sun and moon on the Earth was shown to perturb the Earth's axis
of rotation by calculable amounts; thus, the periodic change in the
direction of the Earth's axis of rotation, an observational fact known
at least since Greek times, was shown to be a mathematical consequence of the law of gravitation.

Lagrange made another notable step in his mathematical analysis
of the motions of the moons of Jupiter. The analysis showed that the
observed irregularities there too were an effect of gravitation. All these
results he incorporated in his *Mécanique analytique* (1788), a work
that extended, formalized, and crowned Newton's work on mechanics.
Lagrange had once complained that Newton was a most fortunate man
in that there is but one universe and Newton had already discovered
its mathematical laws. However, Lagrange had the honor of making
apparent to the world the perfection of the Newtonian theory.

Lagrange's deductions from Newton's laws were extended by
Pierre-Simon Laplace (1749–1827), who ranks with Lagrange and was
his contemporary. Laplace was devoted to any mathematical idea that
would help to interpret nature. Actually, Laplace devoted his entire life
to astronomy, and every branch of mathematics he investigated he
intended to apply to astronomy. There is a familiar story that in his
writing he often omitted difficult mathematical steps and said, "It is
easy to see that . . ." The real point of this story is that he was impa-
tient with mathematical details and wanted to get on with the application. His many basic contributions to mathematics were presented
as by-products of his great work in science and were developed by
others.

One of Laplace's spectacular achievements was the proof that the
irregularities in the eccentricities of the elliptical paths of the planets
were periodic. That is, these irregularities would oscillate about fixed

values rather than becoming larger and larger and so disrupting the orderly motions of the heavens. In brief, the universe is stable. This result Laplace proved in his epochal work, the *Mécanique céleste*, which he published in five volumes over a period of twenty-six years (1799–1825). In this crowning work Laplace summarized the scope of his and Lagrange's results:

> We have given, in the first part of this work, the general principles of the equilibrium and motion of bodies. The application of these principles to the motions of the heavenly bodies had conducted us, by geometrical reasoning, without any hypothesis, to the law of universal attraction, the action of gravity and the motion of projectiles being particular cases of this law. We have taken into consideration a system of bodies subjected to this great law of nature and have obtained, by a singular analysis, the general expressions of their motions, of their figures, and of the oscillations of the fluids which cover them. From these expressions we have deduced all the known phenomena of the flow and ebb of the tide; the variations of the degrees and the force of gravity at the surface of the earth, the precession of the equinoxes; the libration of the moon; and the figure and rotation of Saturn's rings. We have also pointed out the cause that these rings remain permanently in the plane of the equator of Saturn. Moreover we have deduced, from the same theory of gravity, the principal equations of the motions of the planets, particularly those of Jupiter and Saturn, whose great inequalities have a period of above 900 years.

Laplace concluded that nature ordered the celestial machine "for an eternal duration, upon the same principles which prevail so admirably upon the earth, for the preservation of individuals and for the perpetuity of the species."

However, the Newtonian theory of gravitation was to achieve even more astonishing results. One remarkable deduction from the general astronomical theory of Lagrange and Laplace is especially worth mentioning. This was the purely theoretical prediction of the existence and location of the planet Neptune. Galileo saw it in 1613, but he thought it was a star. It had been conjectured that unexplained aberrations in the motion of the planet Uranus, which were observed about 1820, resulted from the gravitational pull on Uranus of an unknown planet. Two astronomers, John Couch Adams (1819–1892), a twenty-six-year-old mathematician at Cambridge, and U. J. J. Leverrier (1811–1877), director of the Paris Observatory in France, working independently, used the observed irregularities and the general astronomical theory to calculate the orbit of the supposed planet. In

1841 Adams calculated the mass, path, and position of what was later called Neptune. He visited Sir George Airy, director of the Royal Observatory at Greenwich, England, to communicate his results. Airy was at dinner, so Adams had to leave his work for Airy to read. Airy did finally read it but was not impressed. In the meantime Leverrier sent directions for locating the planet to the German astronomer John Galle. He observed Neptune that very evening of September 23, 1846. It was barely observable with the telescopes of those days and would hardly have been noticed if astronomers had not been looking for it at the predicted location.

The problem Adams and Leverrier solved was an extremely difficult one because they had to work backward, so to speak. Instead of calculating the effects of a planet whose mass and path were known, they had to deduce the mass and path of the unknown planet from its effects on the motion of Uranus. Their success was therefore regarded as a great triumph of theory and widely proclaimed as final proof of the universal application of Newton's law of gravitation.

The work of Galileo, Newton, and their successors illustrates superbly how our knowledge of the external world is obtained not by sense perceptions but by mathematics. Of course, some observation of bodies falling to earth and some observations of the heavenly bodies suggested the mathematical problems; however, the essence of all the work was mathematical and based primarily on Newton's law of gravitation.

However, all attempts to understand the physical action of the force of gravity failed. Galileo had already questioned the physical nature of gravity. In his *Dialogue on the Great World Systems* he has one of the characters, Salviatus, say: "If he will but assure me, who is the mover of one of those movables [Mars and Jupiter], I will undertake to be able to tell him who maketh the Earth to move. Nay, more: I will undertake to be able to do the same if he can but tell me, who moveth the parts of the Earth downwards." To Salviatus another character, Simplicio, replies: "The cause of this is most manifest, and everyone knows that it is gravity," to which Salviatus responds:

You should say that everyone knows that it is *called* gravity; but I do not question you about the name, but about the essence of the thing . . . not as if we really understood any more, what principle or virtue that is, which moveth a stone downwards, than we know who moveth it upwards, when it is separated from the thrower, or who moveth the moon round, except only the name, which more particularly and properly we have assigned to all motion of descent, namely, gravity.

Newton faced the problem of explaining the action of gravity and said:

> So far I have explained the phenomena of the heavens and of the sea by the force of gravity. . . . I have not yet been able to deduce from the phenomena the reasons for these properties of gravity and I invent no hypotheses [*hypotheses no fingo*]. For everything which is not deduced from the phenomena should be called an hypothesis, and hypotheses, whether metaphysical or physical, whether occult qualities or mechanical, have no place in experimental philosophy. In this philosophy propositions are deduced from phenomena and rendered general by induction. . . . It is enough that gravity really exists, that it acts according to the laws we have set out and that it suffices for all the movements of the heavenly bodies and of the sea.

Newton hoped that the nature of this force would be investigated and mastered. In lieu of this, Newton did have a quantitative formulation of how gravity acted, and this was significant and usable. This is why he says at the beginning of his *Principia:* "For I have design only to give a mathematical notion of the forces, without considering their physical causes and seats." Toward the end of the book he repeats this thought:

> But our purpose is only to trace out the quantity and properties of this force from the phenomena, and to apply what we discover in some simple cases as principles, by which in a mathematical way, we may estimate the effects thereof in more involved cases; . . . We said in a *mathematical way* [the italics are Newton's] to avoid all questions about the nature or quality of this force, which we would not be understood to determine by any hypothesis.

Moreover, in one of his letters to the classical scholar and divine Richard Bentley, Newton expressed the limitations of his success:

> That one body may act upon another at a distance through a vacuum without the mediation of anything else, by and through which their action and force may be conveyed from one to another, is to me so great an absurdity that, I believe, no man who has in philosophic matters a competent faculty of thinking could ever fall into it.

Newton saw clearly that his universal law of gravitation is a description, not an explanation.

Newton said in a second letter to Richard Bentley,

> You sometimes speak of gravity as essential and inherent in matter. Pray, do not ascribe that notion to me; for the cause of gravity is what I do not pretend to know, and therefore would take more time to consider it.

Newton made many statements about gravity in the three editions of his *Mathematical Principles,* but the above are the most specific. Just how gravitation could reach out 93 million miles and pull the Earth toward the sun seemed inexplicable to him, and he framed no hypotheses concerning it. He hoped that others would study the nature of this force. People did try to explain it in terms of pressure exerted by some intervening medium and by other processes, all of which proved unsatisfactory. Later all such attempts were abandoned, and gravitation was accepted as a commonly held but unintelligible fact. Despite total ignorance about the physical nature of gravity, however, Newton did have a quantitative formulation of how it acted, and this was both significant and effective. The paradox of modern science is that although it is content with seeking so little, it accomplishes so much.

The abandonment of physical mechanism in favor of mathematical description shocked even great scientists. Huygens regarded the idea of gravitation as "absurd," because its action through empty space precluded any mechanism. He expressed his surprise that Newton should have taken the trouble to make such a number of laborious calculations with no foundation but the mathematical principle of gravitation. Many others objected to the purely mathematical account of gravitation. The German philosopher and mathematician Baron Gottfried von Leibniz (1646–1716), among others of Newton's contemporaries, criticized his work on this account, holding that his famous formula for the gravitational force is merely a rule of computation, not worthy of being called a law of nature. It was compared adversely with existing "laws," and with Aristotle's animistic explanation of the stone's fall to earth as due to its "desire" to return to its natural place on the ground.

Contrary to popular belief, no one has ever explained the physical reality of the force of gravitation. It is a fiction suggested by the human ability to exert force. The greatest science fiction stories are in the science of physics. However, mathematical deductions from the quantitative law proved so effective that this procedure has been accepted as an integral part of physical science. What science has done, then, is to sacrifice physical intelligibility for the sake of mathematical description and mathematical prediction. Moreover, it has become more and more true since Newton's days that our best knowledge of the physical world is mathematical knowledge. The insurgent seventeenth century found a qualitative world whose study was aided by mathematical abstractions. It bequeathed a mathematical, quantitative world that subsumed under its mathematical laws the concreteness of the physical world.

In Newton's time and for two hundred years afterwards, physicists spoke of the action of gravity as "action at a distance," a meaningless phrase that was accepted as a substitute for explaining the physical mechanism, much as we speak of spirits or ghosts to explain unseen phenomena.

The inability to comprehend the mechanism of gravity accentuates the power of mathematics, for Newton's work was, as the title of his *Mathematical Principles* indicates, entirely mathematical. His work and the additions made by his successors not only provided the calculation of the planetary motions that transcended observations but also enabled astronomers to predict phenomena such as eclipses of the sun and moon to a fraction of a second.

In contrast to the many disorderly and often disastrous events on Earth, the heavenly bodies follow mathematically precise patterns. How did this orderly set of motions come about? Was it likely to continue, or would the Earth someday crash into the sun? Newton's answer was that the universe was the design and handiwork of a divine Creator who would ensure the continuing orderliness. Most eloquent is Newton's statement of the classic argument for the existence of God. In his *Opticks* of 1704 he says:

The main business of natural philosophy is to argue from phenomena without feigning hypotheses, and to deduce causes from effects, till we come to the very first cause, which certainly is not mechanical. . . . What is there in places almost empty of matter, and whence is it that the sun and planets gravitate towards one another, without dense matter between them? Whence is it that nature doth nothing in vain; and whence arises all that order and beauty we see in the world? To what end are comets, and whence is it that planets move all one and the same way in orbs concentric, while comets move all manner of ways in orbs very eccentric, and what hinders the fixed stars from falling upon one another? How came the bodies of animals to be contrived with so much art, and for what ends were their several parts? Was the eye contrived without skill in optics, or the ear without knowledge of sounds? How do the motions of the body follow from the will, and whence is the instinct in animals? . . . And these things being rightly dispatched, does it not appear from phenomena that there is a being incorporeal, living, intelligent, omnipresent, who, in infinite space, as it were in his sensory, sees the things themselves intimately, and thoroughly perceives them; and comprehends them wholly by their immediate presence to himself?

In his second edition of his *Principles,* Newton answers his own questions: "This most beautiful system of sun, planets, and comets could only proceed from the counsel and dominion of an intelligent and powerful Being. . . . This Being governs all things, not as the soul of world, but Lord over all."

In one of his letters to Richard Bentley, Newton repeated this thought:

> To make this [solar] system, therefore, with all its motions, required a cause which understood, and compared together the quantities of matter in the several bodies of sun and planets, and the gravitating powers resulting from thence, the several distances of the primary planets from the sun, and of the secondary ones [moons] from Saturn, Jupiter and the earth; and the velocities with which these planets could revolve about those quantities of matter in the central bodies; and to compare and adjust all these things together in so great a variety of bodies, argues that cause to be not blind and fortuitous, but very skilled in mechanics and geometry.

Newton regarded his homage to God and thereby to theology his greatest contribution.

The vital implications of Galileo's and Newton's work brushed away some of the mysticism and superstition that veiled the heavens, enabling humanity to view them in a more rational light. Newton's law of gravitation cleared out the cobwebs, for it showed that the planets follow the same pattern of behavior as do the familiar objects moving on the Earth. This fact provided additional and overwhelming evidence for the conclusion that the planets are composed of ordinary matter. The identification of the stuff of heaven with the crust of Earth wiped out libraries of doctrines on the nature of heavenly bodies. In particular, the distinction made by the Greek and medieval thinkers between the perfect, unchangeable, and incorruptible heavens and the decaying, imperfect Earth was now shown all the more clearly to be a figment of the human imagination.

It was Newton's work that presented humanity with a new world order, a universe controlled throughout by a few universal mathematical laws, which in turn were deduced from a common set of mathematically expressible physical principles. Here was a majestic scheme that embraced the fall of a stone, the tides of the oceans, the motions of the planets and their moons, the defiant sweep of comets, and the brilliant, stately motion of the canopy of stars. The Newtonian scheme was decisive in convincing the world that nature is mathematically designed and that the true laws of nature are mathematical.

The work of Copernicus, Kepler, Galileo, and Newton made possible the realization of many dreams. There was the dream and hope of ancient and medieval astrologers of anticipating nature's ways. There was also the plan that Bacon and Descartes had advanced of mastering nature for the improvement of human welfare. Humanity progressed toward both these goals: the scientific and the technological. The universal laws certainly made possible the prediction of phenomena they comprised, and mastery is but a step away from prediction, for knowing the unfailing course of nature makes possible the employment of nature in engineering devices.

Still another program for probing and understanding nature found fulfillment in the work of Galileo and Newton. The Pythagorean–Platonic philosophy that number relations are the key to the universe, that all things are known through number, is an essential element in the Galilean scheme of relating quantitative aspects of phenomena through formulas. This philosophy was kept alive throughout the Middle Ages, although most often, as with the Pythagoreans themselves, it was part of some larger mystical theory of creation, with number as the form and cause of all created objects. Galileo and Newton divested the Pythagorean doctrine of all its mystical associations and reclothed it in a style that set the fashion for modern science.

Man today uses the Newtonian theory to send people to the moon, to send spaceships to photograph planets such as Mars and Saturn, and to launch satellites that circle the Earth (an idea that had occurred to Newton). All of the planning based on the mathematical theory works perfectly. Any misadventures would result from the failure of human mechanisms.

VII

Mathematics and the Imperceptible Electromagnetic World

There are more things in heaven and earth, Horatio, than are dreamt of in your philosophy.   Shakespeare

We have seen some examples of how the seventeenth- and eighteenth-century mathematicians and physicists, starting from phenomena that were perceived by the senses, as in astronomical and terrestrial motions, built up superb mathematical theories that extended human knowledge of these phenomena, corrected and explained some illusions, and gave us some understanding of what we believed to be the design and operation of nature. Theories, similar in most respects, were also promulgated in the areas of heat, hydrodynamics (the behavior of fluids and gases), and elasticity. Of all of these contributions one could cite the precept of Aristotle that there is nothing in the intellect that is not first in the senses. Of course, the mathematical theories transcended observations and even introduced, as in the case of gravity, concepts that have no apparent reality. Nevertheless, the predictions made on the basis of the theories agreed excellently with experience. One might say that experience is merely a corroboration of these theories.

It was true that, despite the belief that nature was a vast mechanism, scientists had failed to discover and explain the modus operandi of gravity and light. In the latter case, the belief in ether served to quench any doubts about whether there was any mechanism, even though the details were yet to be supplied. As for gravity, the nature of its operation was entirely unknown. However, the successes of Newton, Euler, d'Alembert, Lagrange, and Laplace in describing and predicting mathematically an enormous variety of astronomical phenomena were so remarkably accurate that scientists were thrilled and even smug about their successes. They lost sight of the absence of a physical mechanism and concentrated on the mathematics. Laplace did not question the appropriateness of the title of his five-volume classic, *Celestial Mechanics* (1799–1825).

Developments of the nineteenth and twentieth centuries, which we are about to relate, raised fundamental questions about the nature and contents of our physical world. The first of these developments, dealing with electricity and magnetism, added another phenomenon to our physical universe. Like the discovery of Neptune, this one, too, could hardly have been made without the aid of mathematics. Unlike Neptune, however, this addition was decidedly insubstantial. It weighed nothing, and it could not be seen, touched, tasted, or smelled; it was and is physically unknown to us. Moreover, unlike Neptune, this shadowy substance has had manifest and even revolutionary effects on the daily lives of nearly every man, woman, and child in contemporary civilization. It whisks communications around the world in the flicker of an eyelid; it extends the political community from the neighborhood street corner to all of planet Earth; it quickens the tempo of life, promotes the spread of education, creates new arts and industries, and revolutionizes warfare. Indeed, hardly a phase of human life has been unaffected by the science of electromagnetism.

As in the cases of astronomy, sound, and light, our knowledge of electricity and magnetism begins with the Greeks. Thales (*c.* 640– *c.* 546 B.C.) knew that iron ore containing lodestone, such as that found near Magnesia in Asia Minor, attracts iron. The medieval Europeans learned from the Chinese that if a piece of lodestone is allowed to rotate freely it will orient itself along roughly the North–South direction and hence can be used as a compass. Thales is also credited with knowing that if amber is rubbed it attracts light particles of matter such as straw. This is the beginning of the science of electricity, the Greek word for amber.

The first serious study of magnetism was made by William Gilbert (1540–1603), court physician to Queen Elizabeth. His *On the Magnet and Magnetic Bodies and on the Great Magnet, the Earth* (1600) is still a clearly readable account of simple experiments that proved, among other things, that the Earth itself is a huge magnet. Gilbert also found that there are two kinds of magnetism, North-seeking and South-seeking, or more simply North and South, often labeled positive and negative, respectively. Two positive or two negative magnetic substances repel each other, whereas oppositely endowed magnetic substances attract each other. These opposite types of polarities, as they are called, are found, for example, on the opposite ends of a bar magnet. Moreover, the peculiar property of magnets is their power to attract unmagnetized iron or steel, a stronger magnet being able to pull a heavier piece of iron to itself.

Gilbert also explored the second phenomenon observed by Thales, namely, the electrification of rubbed amber. He found that sealing wax when rubbed with fur, or glass when rubbed with silk, also acquire the property of attracting light particles of matter. These experiments suggested that there are two kinds of electricity. Moreover, as in the case of magnetism, two objects possessing the same kind of electricity repel each other, and objects possessing different kinds attract each other. However, Gilbert made little progress in understanding the physical nature of magnetism or electricity.

Gilbert did recognize that there is an essential difference between magnets and electric charges. We can rub glass with silk and cause the glass to be positively charged and the silk negatively charged. We can then separate the glass and silk and have at our command a positive charge on the glass that is completely independent of the negative charge on the silk. While there are likewise two kinds of magnetism, positive and negative, or North and South, and while, as in the case of electric charges, the opposite types attract each other and like types repel each other, the two magnetic types cannot be separated in any physical objects.

However, a long series of subsequent investigations whose details we need not follow showed that this account of electricity is not correct. During most of this century physicists believed that there is just one kind of electricity.* They had discovered that there are tiny bits of matter, indeed the smallest pieces found in nature, and they called them electrons. We cannot see electrons any more than we can see the larger bits of matter called atoms that contain the electrons, but the indirect evidence for their existence became quite strong. An object that is negatively charged—that is, one that exhibits the behavior of rubbed silk—contains an excess of electrons. On the other hand, objects that were formerly described as being positively charged, such as glass after being rubbed with silk, were recognized to be deficient in electrons. Apparently rubbing the glass with silk loosens a number of

---

*Electricity had been regarded by many physicists as a fluid, and by others as consisting of two fluids, until about 1900, when the electron theory was generally adopted. (However, see Chapter X for further developments.)

electrons from the glass, and these attach themselves to the atoms of the silk. Hence the glass, deficient in electrons, becomes positively charged and the silk negatively charged. An object that contains the normal number of electrons is said to be electrically neutral.

With the proper experimental setup, the action of electrically charged bodies can be studied. For example, if two little glass balls, positively charged, are hung on strings and placed alongside each other, the balls will repel each other because both are positively charged. Because electrically charged objects exert forces on each other, and because magnetic poles do likewise, it is evident that in the phenomena of electricity and magnetism we have forces at our disposal that can be studied to advantage. Let us investigate first the behavior of electricity.

The scientists of the late eighteenth century who became engrossed in the study of the forces exerted by electrically charged bodies had learned the lesson taught by Galileo and Newton, that is, to seek basic quantitative laws. The first law they discovered was indeed a surprise to them. Because the force exerted by one electrified object on another depends on the quantity of electricity in each, it is necessary to have a measure of this electricity. Hence some quantity is chosen as a standard (just as some quantity of mass is chosen as a unit of mass), and the quantity of electricity in any object is then measured in terms of that standard. One of the commonly used units of charge is called the coulomb, after the French physicist Charles Augustin Coulomb (1736–1806), who discovered the law of force we are about to describe. If one has two quantities of charge, q_1 and q_2, these will attract or repel each other depending on whether they are unlike or like. Coulomb found the remarkable law that the force F of attraction or repulsion is given by the formula

$$F = k \left(\frac{q_1 q_2}{r^2} \right)$$

where r is the distance between the two collections of charges q_1 and q_2, and k is a constant. The value of k depends on the units used to measure charge, distance, and force.

The remarkable fact about the formula is its identity with the form of the law of gravitation. The charges q_1 and q_2 act like masses, and the force varies inversely with distance in exactly the same way as the force of gravitation acts between two masses. Of course, the electrical force can be attractive or repulsive, whereas the gravitational force is always attractive.

In the late eighteenth century Professor Luigi Galvani (1737–1798) formed a piece of wire consisting of two different metals and then inserted the two ends of the entire wire into the nerve of a frog's leg. The frog's leg twitched. Galvani, who had been studying animal electricity, attributed the twitch to some electrical current in the frog. However, the significance of this discovery was appreciated by another Italian, Alessandro Volta (1745–1827), a professor of physics at the University of Padua. Volta realized that the two unlike metals were producing a force, now called electromotive force, at the ends of the wire, and he worked out a more effective combination of metals, that is, a battery. By replacing the frog's nerve with a wire and by attaching the ends of the wire to the terminals of his battery, Volta showed that the force could be utilized to make minute particles of matter flow in the wire. This flow of particles, later identified as electrons, is an electric current. Volta's battery made these electrons flow instead of leaving them bunched up and stationary as they are on rubbed amber. Incidentally, Volta's battery is not different in principle from the modern automobile and flashlight batteries. The strength of batteries is now measured in volts in honor of Volta, and current is measured in amperes in honor of a man we shall meet shortly. One ampere is one coulomb per second or 6×10^{18} electrons per second.

Thus far, electricity and magnetism were considered to be distinct or unrelated phenomena. However, this situation was to change radically, and the relationship that was discovered will bring us to the heart of our story. The first important relationship was discovered in 1820 by the Danish physicist Hans Christian Oersted (1777–1851), a professor of natural philosophy at the University of Copenhagen. Using Volta's idea of a battery to run electric current through a wire, Oersted found that the current deflects a compass needle placed over the wire. When the direction of the electric current is reversed, the needle reverses direction. Another way of describing Oersted's discovery is to say that the electric current sets up a magnetic field about the wire. This magnetic field attracts and repels other magnets, as does the natural lodestone.

The next fundamental relationship between electricity and magnetism was discovered by the French physicist André-Marie Ampère (1775–1836), a professor at the Ecole Polytechnique who had heard about Oersted's work. In 1821 Ampère found that two parallel wires carrying electric currents also behave like two magnets. If the currents are in the same direction, the wires attract each other; and if in opposite directions, they repel each other.

It remained for a self-educated, ex-bookbinder's apprentice, Michael Faraday (1791–1867), who was working in England, and a schoolmaster, Joseph Henry (1797–1878), of the Albany Academy in New York State, to discover the other essential link between electricity and magnetism and thereby set the stage for the dramatic entrance of Maxwell. If a wire carrying current sets up a magnetic field, will not a magnetic field induce current in a wire? The answer, as these men showed in 1831, is yes, provided that the wire is present in a varying magnetic field. This phenomenon is called electromagnetic induction.

Let us examine more closely the essence of Faraday's and Henry's discovery. Suppose that a rectangular frame of wire (Figure 28) is rigidly attached to a rod R, and that the frame and rod are then placed in the field of a magnet. When the rod is made to rotate—by the use of water power or a steam engine, say—the frame of wire, which is rigidly attached, will also rotate. Suppose, too, that the rod (which is insulated from the wire) rotates at a constant speed in a counterclockwise direction and that the wire BC starts from its lowest position. As BC goes from this position toward a horizontal position on the right, a flow of electric current takes place throughout the rectangle of wire in the direction from C to B. This flow reaches a maximum at that horizontal position. As BC continues upward, the flow decreases in quantity, and it vanishes when BC is at the highest position. As BC continues to rotate, a current again appears in the wire, this time in the direction from B to C. Again, the flow increases in quantity as the wire rotates and reaches a maximum value for the new direction of flow when BC is again horizontal. As BC returns to the lowest position of its path,

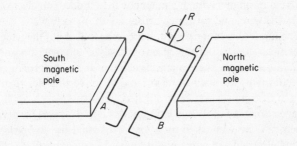

FIGURE 28

the flow of current diminishes and finally disappears. This cycle of changes repeats itself with each complete rotation of the rod. The appearance and flow of current in a wire that is moved in the field of a magnet are other examples of the phenomenon of electromagnetic induction.

The current generated is a flow of billions of minute, invisible particles of matter called electrons. The amount of the flow created by electromagnetic induction varies with time and, because we are dealing with measurable quantities, we can formulate the functional relation involved. The relation between current and time is certainly periodic in that the sequence of variations repeats itself with each complete rotation of the wire frame. It may be too much to expect that in this periodic phenomenon, the function sine x should serve. Yet nature never ceases to accommodate itself to man's mathematics. The relation between current I and time t is of the form

$$I = a \sin bt$$

where the amplitude a depends on such factors as the strength of the magnet, and the frequency b depends on how fast the frame rotates. If it makes 60 rotations in 1 second, then the value of b is 60×360 or 21,600. (The function $y = \sin x$ goes through one cycle in 360 degrees. Hence a 60-cycle current will go through 60×360 degrees in 1 second. If the current flows for t seconds, the number of degrees is $60 \times 360t$.) The current that furnishes electricity to most homes goes through 60 complete sinusoidal cycles of change in 1 second; for this reason it is called 60-cycle alternating current.

Electric current, then, can be represented by a mathematical formula. But how does the process of electromagnetic induction produce electric currents? This phenomenon is replete with mystery. Somehow the mere motion of a wire in a magnetic field induces an electromotive force in the wire, and this force causes a current to flow.

The modern reader does not have to be told about the widespread uses of electricity and of the effect of this type of power on our civilization, but it may be well to emphasize that the principles of generating electricity by mechanical means and of converting electrical power to mechanical power were investigated long before such applications were even dreamed of. While Faraday was doing his early experiments in electricity, he was asked by a visitor of what use his principle of inducing electricity in wires might be, and Faraday replied, "What is the use of a newborn child? It grows to be a man." On another occasion

Faraday was visited by Mr. Gladstone when he was chancellor of the exchequer. Gladstone asked the same question, and this time Faraday answered, "Why, sir, presently you will be able to tax it."

Faraday extended our knowledge of electromagnetic induction by another significant experiment. He placed two coils of wire near each other as shown in Figure 29. His plan was to make a current flow in the left-hand coil *CD* and have this current set up a magnetic field whose direction is shown by the oval lines in the figure. This magnetic field would extend to the wire in the second coil *EF*. However, Faraday wanted a changing magnetic field; so he connected the terminals *A* and *B* of the first coil to a source of alternating current. This alternating current passes through the coil and sets up a varying magnetic field through and about it in accordance with Oersted's principle. Thus, as the alternating current increases in strength, a stronger magnetic field appears around the coil *CD,* and as it decreases in strength the magnetic field decreases. Because the coil *EF* is alongside the coil *CD,* the magnetic field generated by the current in the coil *CD* surges through the coil *EF.*

Thus, Faraday had produced a changing magnetic field that moves past a wire, the coil *EF.* If a magnetic field moving past a fixed wire produces a force in the wire, this magnetic field should set up a force or voltage and therefore a current in the coil *EF.* Moreover, because the magnetic field not only moves through the coil *EF* but increases and decreases in strength, the current induced in the coil *EF* should do likewise. That is, the current in the coil *EF* should be alternating. Faraday further expected that this current should persist as long as the alternating current was maintained in the first coil and hence that he would be able to study the induced current at length.

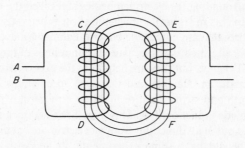

FIGURE 29

Faraday did find that an alternating current appeared in the coil *EF;* moreover, as he had expected, the frequency of this current was precisely the frequency of the current applied at the terminals *A* and *B* of the first coil. One apparent use of this principle is to transfer electric current from one coil to another, even though the second one is not connected to the first one. This use is made in what are now called voltage transformers, but we shall not pursue this application because it would take us far afield from our major concern.

With Faraday's discovery of the important principle of electromagnetic induction, and thereby of a new link between magnetism and electricity, the science of electromagnetism—the word used to denote the interaction of electricity and magnetism—now had a number of significant advances to its credit. But the phenomena were becoming more involved, and Faraday was beginning to have difficulty in thinking about them. In the cases of simple electric and magnetic fields it had been rather easy to construct some physical pictures and to obtain, either by measurement or simple reasoning, the appropriate mathematical laws. In the case of electromagnetic induction, the problem of determining the electric force and current induced in the second coil, knowing the current in the first one, was already too complex to be easily analyzed. This phenomenon involved, first of all, calculating the strength of the magnetic field that accompanied the current in the first coil and then calculating the voltage and current induced in the second coil. Moreover, because he had discovered what seemed to be a promising physical process, Faraday wished to know just how he could increase its effectiveness. Would more current in the first coil or a longer coil or a wider coil produce more current in the second one? How should the coils be placed relative to each other?

Faraday did conclude that the magnetic effects of electricity are propagated by means of actions among the contiguous particles of the medium surrounding the electrified bodies and called the medium the dielectric. In this medium the magnetic effect is activated by lines of magnetic force that are not visible but that Faraday believed were real.

Faraday admitted that speculations about lines of force are liable to error and change, but they are helpful to the experimentalist and to the mathematician. He also said that such speculations lead to real physical truth and strove hard to build up a physical explanation of electromagnetic induction. It was his hypothesis that magnetic lines of force spread out in all directions from an electric current or a magnetic pole, and he did have some experimental evidence to support his attempted physical explanation: for example, if iron filings are dropped in the space around a bar magnet, they automatically line up along the lines of force.

Although Faraday was well aware of what mathematics could accomplish, his genius was confined to experimentation and physical thinking. In the case of complicated electromagnetic phenomena, physical thinking is at a considerable disadvantage. It is easy to visualize the motion of a projectile, its angle of fire, and its range. Electric and magnetic fields, on the other hand, are not visible, and their configurations are not readily obtained. Despite his past successes in concocting physical pictures, Faraday realized that physical thinking was not going to carry him much further. Faraday was at that stage where the physics becomes too difficult for the physicists and requires the services of a mathematician.

Fortunately, the great nineteenth-century mathematical physicist James Clerk Maxwell (1831–1879) was assiduously preparing himself for the task. As a youth Maxwell showed promise of being able to make first-class contributions. A paper he wrote at the age of fifteen on mechanical methods of generating some curves was published in the *Proceedings of the Royal Society of Edinburgh*. During his attendance at the universities of Edinburgh and Cambridge his professors and fellow students recognized his brilliance and originality. In 1856 he was chosen to be professor of physics at Marischal College in Aberdeen. A few years later he transferred to King's College in London and in 1871 moved to Cambridge University.

Like all scientists, Maxwell took up the challenging problems of his day. He invented color photography and was one of the formulators of the kinetic theory of gases. However, our concern here is with his work in electromagnetism. He sought to synthesize in one theory all the known phenomena of electricity and magnetism. He began his work in electromagnetism by reading Faraday's *Experimental Researches*. In 1855, at the age of twenty-three, he published his first paper on the subject: "On Faraday's Lines of Force." In this and later papers Maxwell undertook the task of translating Faraday's physical researches into mathematical form.

In the early 1850s Maxwell was very much influenced by the work of William Thomson (Lord Kelvin; 1824–1907). Thomson favored a mechanical explanation of electric and magnetic phenomena, and he used fluid flow, the flow of heat, and elasticity as a guide. He applied these analogies to ether, which he regarded as a field wherein forces act between contiguous particles—a concept suggested earlier by the mathematicians Cauchy, Poisson, and Navier—as opposed to action at a distance. Maxwell also sought some mechanical explanation of the action of ether. However, neither he nor Thomson was successful. Nevertheless, Thomson had introduced what is now called the field concept, as opposed to action at a distance, and Maxwell adopted this. Thomson had also made a start on the mathematics of wave propagation, and Maxwell benefited somewhat by this.

Maxwell obtained new insights into the unresolved problem of electromagnetic induction in 1861, using the notion of the ether as an elastic medium. It was apparent from Faraday's work on the transmission of electric current from one coil of wire to another that magnetic fields can travel some distance. Maxwell also concluded that a varying electric current penetrates the space surrounding the first coil of wire. This current, which he called a displacement current, would account for some effects of electric currents quite distant from the actual physical currents in a wire. In this paper Maxwell noted his first glimpse of displacement current, but it was still not clear nor complete.

To justify and complete his understanding of displacement current, Maxwell considered the action of a condenser in an electric circuit. A condenser consists of two parallel plates separated by some insulating medium such as air or even a vacuum. Yet an alternating electric current passes from one plate to another. To Maxwell it was evident that the ether transmitted displacement current from one plate to the other.

In 1865 Maxwell published his key paper, "A Dynamical Theory of the Electromagnetic Field," in which he discarded all physical models and presented the appropriate mathematics. His equations contained a new term that physically represented his displacement current. This mathematical formulation convinced him that such currents can travel great distances.

The nature of this displacement current requires some additional explanation. Following Faraday, Maxwell regarded electric and magnetic fields as existing in the space around magnets and around wires carrying currents. Ampère's law itself deals with a current flowing in a wire. However, when this current alternates (for example, suppose it varies sinusoidally in time), the electrons in the wire move rapidly back and forth. Then, the electric field they set up will also move, and at any point in space outside the wire the strength of this electric field will vary with time. Hence, an alternating current in a wire can be considered to be accompanied by a varying electric field in the space surrounding the wire. Maxwell accepted the reality of this varying electric field and observed that it had the mathematical properties of a current, even though (apart from the wire that gave rise to it) it did not in itself consist of the motion of electrons. He justified calling this varying electric field a displacement current, because it amounted to the displacement or variation of an electric field. Maxwell's own words in his *Treatise on Electricity and Magnetism* (1873) clearly describe his conclusion:

> One of the chief peculiarities of this treatise is the doctrine which it asserts, that the true electric current, that on which the electromagnetic phenomena depend, is not the same thing as the current of conduction [the current in the wire], but that the time variation of the electric displacement must be taken into account in estimating the total movement of electricity.

Maxwell pursued the mathematical implications of the existence of displacement current. Oersted's law said that a current in a wire is accompanied by a magnetic field. But since Maxwell had added the displacement current to the conduction current, or current in the wire, he concluded that the displacement current also gave rise to a magnetic field, and this magnetic field was part of the one formerly regarded as produced by the conducting current alone. In other words, the magnetic field surrounding the wire must result from both currents, conduction and displacement.

To recapitulate, the first bold step Maxwell made was to introduce the concept of a displacement current and to speculate that this current, which existed in space rather than in wire, also gave rise to a magnetic field. Thus, he revised Ampère's law to relate the total current—that is, conduction and displacement—to the magnetic field emanating from the wire. Hence, a most important part of Maxwell's law is that a changing electric field, whether arising in a conduction current or in a displacement current, creates a magnetic field. If we now recall that Faraday's law, as formulated by Maxwell, states that a changing magnetic field creates a varying electric field, we see that Maxwell introduced the reciprocal relationship.

We are in a position now to understand what Maxwell predicted from mathematical reasoning. Waves originating in the sinusoidal current in the coil *CD* (Figure 29) give rise to the varying electric field in the space around it, which creates the varying magnetic field. But the magnetic field creates a varying electric field and this, in turn, a varying magnetic field. What will these fields do under the constant "pressure" exerted by the current in the coil *CD?* The answer is almost obvious. They will spread out into space and reach points far removed from the coil *CD*. They might even reach a *distant* coil *EF*. There, the changing electric field will cause a current to flow in the wire, and this current can be used for any purposes currents serve. Thus, Maxwell discovered that an electromagnetic field—that is, a combination of a changing electric and magnetic field—will travel *far out into space*. Incidentally, Faraday had already suspected that this might be possible when he considered what would happen if the coil *EF* were somewhat removed from the coil *CD*. But what Faraday conjectured on physical grounds, without comprehending the mechanism or recognizing the existence of displacement currents, Maxwell established on mathematical grounds.

Waves have a wavelength and a frequency of variation per second. In the case of electromagnetic waves, the wavelength is determined (although this may not be immediately obvious) by the size of the coil used. To keep the coils (or whatever wires are used to send electromagnetic waves into space) reasonably small, it is necessary that the wavelength be small.

To explain these quantities let us consider a sinusoidal wave that has the characteristics shown in Figure 30. One cycle is the graph from *O* to *A*. This cycle repeats itself many times a second, and the number of such cycles per second is the frequency of the cycles. What is called the wavelength λ (lambda) is the distance from *P* to *Q*. The distance the wave travels per second is the wavelength times the frequency *f*, giving the formula:

$$\lambda f = c$$

where *c* is the velocity of the wave motion.

Electromagnetic waves are a bit more complicated. Not only does the electric field move out sinusoidally, so also does the magnetic field.

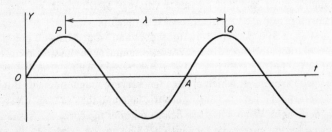

FIGURE 30

Moreover, these two fields are at right angles to each other, and both are at right angles to the direction in which the field moves. Figure 31 shows that the directions of the electric field E and the magnetic field H are at right angles to each other.

Thus, Maxwell's first and greatest discovery is that electromagnetic waves can travel thousands of miles from a source and presumably be detected by some suitable apparatus at a distant point. In the course of his mathematical work, Maxwell made a second sensational discovery, this one concerning light. The phenomenon of light had been studied from Greek times onward, and after numerous experiments, two physical theories competed for an explanation. One maintained that light consists of tiny, invisible particles that move along rays. The other theory was that light is a motion of waves, and various explanations of how these waves were formed and moved through space were offered. Both theories explained somewhat satisfactorily the reflection of light and refraction, that is, the change in direction when passing, say, from air to water. However, the diffraction of light—that is, the bending of light around an obstacle such as a disk—was more reasonably explained by a wave theory such as that which accounts for water waves bending around the stern of ship. During the early part of the nineteenth century Thomas Young (1773–1829) and Augustin Fresnel (1788–1827) argued effectively for the wave theory involving a medium they did not specify.

One more early development in the science of light is relevant. In 1676 the Danish astronomer Olaus Roemer (1644–1710) showed that the velocity of light is finite and obtained a good approximation to that velocity, namely, 2.2×10^{10} centimeters per second. He obtained this figure by measuring the difference in the time that an eclipse of Jupiter by one of its moons is observed when Earth is moving away from Jupi-

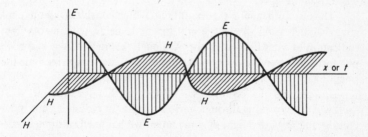

FIGURE 31

ter as against when Earth was moving toward Jupiter. The difference in distance traveled by the light from Jupiter's moon was about the diameter of Earth, and he measured the time difference. More accurate measurements in the nineteenth century showed that the velocity of light is about 186,000 miles per second.

In the course of his mathematical work Maxwell found that the velocity of electromagnetic waves is 186,000 miles per second. Now, the velocity of light, as calculated by Olaus Roemer and later physicists, was known to be about the same figure. The identity of these velocities and the fact that electromagnetic radiation and light were both wave motions induced Maxwell to declare that light is an electromagnetic phenomenon. In 1862 Maxwell said, "We can scarcely avoid the inference that light consists in the transverse undulations of the same medium [ether] which is the cause of electric and magnetic phenomena." He wrote a paper on this in 1868. Maxwell's inference became and remains the current theory of light. (See also Chapter IX on relativity.)

Maxwell's conclusion that light is an electromagnetic phenomenon superseded all of the older work. More accurately, white light (for example, sunlight) is a composite of many frequencies, so that there is an entire spectrum of frequencies in the range of visible light. Specifically, all waves whose frequencies range from 4×10^{14} to 7×10^{14} are visible waves. To our eyes these different frequencies register different colors. As the light received varies from the smallest frequency to the largest in the range given above, our sensations of color, a contribution of the nerves and brain, change gradually from red to yellow, to green, to blue, and finally, to violet. We can combine simple colors to produce new ones. White light itself, for example, is not a simple color "tone," but a light "chord," a composite effect of many colors. Thus, sunlight contains all colors from red to violet, the composite effect of which is white light.

The electromagnetic theory of light, which considers light to consist of a succession of electric and magnetic fields, gave us our best indication of what light might be. Although theories of light had been put forth before Maxwell's work, none adequately explained all the phenomena. The electromagnetic theory of light proved to be satisfactory and granted new power to scientists to predict, for example, what light will do in passing through various media. In particular, the older concept that regarded light as some unknown but fixed substance traveling along rays and subject to the laws of reflection and refraction was seen to be just a good approximation, for, strictly speaking, a light wave that travels out in space is not confined to a set of lines, and its strength varies with time at any given point, and from point to point at any given time; in other words, it behaves just like water waves traveling out from a source. However, the variations are so small and so rapid that light appears to be a constant flow.

The prediction based on mathematical reasoning that light is an electromagnetic wave illustrates one of the remarkable values of mathematics. In the words of the foremost contemporary philosopher Alfred North Whitehead, "The originality of mathematics consists in the fact that in the mathematical sciences connections between things are exhibited which, apart from the agency of human reason, are extremely unobvious."

In Maxwell's time, physicists had already become somewhat informed regarding the existence and properties of ultraviolet (UV) waves, which, though invisible to the eye, make their presence known by blackening photographic film. Moreover, infrared rays, also invisible, convey heat readily registered by a thermometer. Both of these types of rays are present in the sun's radiation. They can also be generated by passing electric current through special filaments, in the same manner that visible light is created by passing a current through tungsten. The conjecture that infrared rays and ultraviolet rays are electromagnetic waves was readily established experimentally, and infrared rays were found to have frequencies a little below those of visible light whereas ultraviolet rays proved to have frequencies a little above.

More and more pieces of the electromagnetic jigsaw puzzle were soon filled in. In 1895 Wilhelm Conrad Roentgen (1845–1923), a German physicist, discovered X rays, which were soon identified as electromagnetic waves with frequencies even higher than those of ultraviolet waves. Finally, gamma (γ) rays, which issue from radioactive substances, were discovered and were also found to be electromagnetic waves with frequencies still higher than those of X rays.

The electromagnetic spectrum varies from wavelengths of 10^{-14} to 10^8, that is, over a range of 10^{22}. In terms of doubling, $10^{22} = 2^{73}$. Of the seventy-three "octaves," the visible band covers only one; hence our eyes are very limited. But we do have instruments to detect infrared, ultraviolet, X, and γ radiation.

What kind of reception could such an "ethereal" theory receive? In 1873 almost all physicists were skeptical of the existence of electromagnetic waves. At the very least, they found the concept hard to comprehend. An exception was Hendrik Antoon Lorentz (1853–1928), who tried unsuccessfully to produce the various types of waves experimentally. However, in his doctoral thesis of 1875 he showed that Maxwell's theory explained reflection and refraction of light better than any other existing theory.

Clearly there was need of experimental confirmation, for mathematical predictions based on physical axioms—in the present case Maxwell's equations—are not certain because physical axioms may be faulty. In 1887, about twenty-five years after Maxwell had predicted the existence of electromagnetic waves, Heinrich Hertz (1857–1894), another famous physicist and brilliant student of Helmholtz, generated electromagnetic waves and received them in a coil some distance away from the source. These waves were called Hertzian waves for a long time and are none other than the radio waves employed today in a thousand ways. This confirmation was startling and was soon followed by various applications.

Soon thereafter, in 1892 Sir William Crookes, a British experimenter, proposed wireless telegraphy. By 1894 Sir Oliver Joseph Lodge, among others, had sent waves over short distances. Finally, Guglielmo Marconi in 1901 conceived the idea of sending electromagnetic waves a great distance by building a special antenna. Telegraphic signals, he thought, could cross the Atlantic from Europe to North America. The transmission of speech soon followed. In 1907 Lee De Forest invented the radio vacuum tube, and the electromagnetic transmission of speech and music became common.

The radio transmission of speech was a remarkable discovery. Sound travels at about 1100 feet per second. If sound waves could reach San Francisco from New York City, we would have to wait about eight hours for the sonic reply to come back. By telephone the reply is immediate, because most of the message is carried by radio waves that travel at 186,000 miles per second.

We use so many forms of electromagnetic waves nowadays that we lose sight of this remarkable feature. Let us consider, for example, the process of television. The variations in brightness of a scene being televised are converted into electric current; this is converted into electromagnetic waves that travel through space; the waves induce electric current in a receiving antenna; from this the current passes through an electric circuit; finally, by means of a cathode-ray tube the current is transformed into light waves.

The propagation of electromagnetic waves through space poses a major problem. Although Maxwell himself had written in 1856, "and a mature theory, in which physical facts will be physically explained, will be formed by those who by interrogating Nature herself can obtain the only true solution of the questions which the mathematical theory suggests," we have not the slightest physical notion of what it is that travels from transmitter to receiver. Despite the Herculean efforts to determine physically what an electric field and a magnetic field are, scientists are unsuccessful.

When Maxwell proved that electromagnetic waves travel with the velocity of light, he concluded that these waves travel in ether, because since Newton's days ether had been accepted as the medium in which light moved. Moreover, because the waves move with enormous velocity, the ether must be highly rigid—for the more rigid a body, the faster waves move through it. On the other hand, if ether pervades space it must be completely transparent and the planets must move through it with no friction. These conditions imposed on ether are contradictory. Moreover, ether cannot be touched, smelled, or isolated from other substances. Such a medium is physically incredible. We must conclude that it is a fiction, a mere word satisfying only those minds that do not look behind words. Furthermore, the entire account in terms of fields is a crutch that helps the human mind to propel itself forward but must not be taken literally or seriously.

To sum up, then, we do not have any physical account of the action of electric and magnetic fields, nor do we have any physical knowledge of the electromagnetic waves as waves. Only when we introduce conductors such as radio antennas in electromagnetic fields do we obtain any evidence that these fields exist. Yet we send radio waves bearing complex messages thousands of miles. Just what substance travels through space we do not know.

It is equally disturbing to realize that these waves are all around us. We have but to put a radio receiver or a television set into operation to receive waves sent out by dozens of radio and television stations; yet our senses have not the slightest perception of their presence.

Our ignorance about the physical nature of electromagnetic waves disturbed many of the principal creators of the theory. William Thomson (Lord Kelvin), in a speech of 1884, was not satisfied with Maxwell's work. He said, "I never satisfy myself until I can make a mechanical model of a thing. If I can make a mechanical model I can understand it. As long as I cannot make a mechanical model all the way through I cannot understand; and that is why I cannot get the electromagnetic theory." What was lacking was a mechanical theory of the ether. Helmholtz and Lord Kelvin dismissed Maxwell's displacement current as a fiction.

Although Maxwell did try unsuccessfully to obtain a mechanical theory of electromagnetic phenomena in terms of pressures and tensions in an elastic medium, and the later efforts of Heinrich Hertz, William Thomson, C. A. Bierknes, and Henri Poincaré were equally unsuccessful, the experimental evidence for Maxwell's theory marked the end of any opposition. The adoption of Maxwell's theory meant also the adoption of a purely mathematical view, for the supposition that an electromagnetic wave consists of a conjoined electric and magnetic field traveling through space hardly explains the physical nature. By embracing in one theory light, X rays, and so forth, it merely reduces the number of scientific mysteries by compounding one of them.

Hertz said, "Maxwell's theory consists of Maxwell's equations." There is no mechanical explanation, and there is no need for one. He continued, "One cannot escape the feeling that these equations have an existence and an intelligence of their own, that they are wiser than we are, wiser even than their discoverers, that we get more out of them than was originally put into them."

The precise and comprehensive account of electromagnetism is the mathematical account. Hence electromagnetic theory is entirely a mathematical theory illustrated by a few crude physical pictures. These pictures are no more than the clothes that dress up the body of mathematics and make it appear presentable in the society of physical sciences. This fact may disturb or elate the mathematical physicist, depending on whether the mathematician or the physicist is dominant.

No one appreciated the thoroughly mathematical character of electromagnetic theory more than Maxwell. Although he had tried almost desperately to build a physical account of electromagnetic phenomena, in his classic *Treatise on Electricity and Magnetism* (1873) he omitted most of this material and emphasized the highly polished, complex mathematical theory. He himself had once advised a preacher who seemed to be speaking over the heads of his congregation, "Why don't you give it to them thinner?" Yet his own attempt to "thin down" the mathematics of electromagnetic fields with an intuitively understandable explanation was unsuccessful. Radio waves and light waves operate in a physical darkness illuminated only for those who would carry the torch of mathematics. Furthermore, whereas it is possible in some branches of physics to fit mathematical theory to physical facts, about the best one can do in electromagnetic theory is to fit inadequate physical pictures to mathematical facts.

Maxwell set the tone and practice of modern mathematical physics. It is primarily mathematical. Maxwell's electromagnetic theory surpasses even Newton's gravitational theory in embracing a variety of seemingly diverse phenomena in one comprehensive set of mathematical laws. The behavior of the grain of sand and the heaviest star can be described and predicted with Newton's laws of motion. The invisible varieties of electromagnetic waves including light can be described and harnessed with Maxwell's electromagnetic laws. Electric currents, magnetic effects, radio waves, infrared waves, light waves, ultraviolet waves, X rays, and gamma rays, sinusoidal waves with frequencies as low as sixty per second and as high as one followed by twenty-four zeros are manifestations of one underlying mathematical scheme. This theory, which is at once so profound and so comprehensive that it beggars the imagination, has revealed a plan and an order in nature that speaks more eloquently to humanity than nature herself.

Electromagnetic theory affords us another illustration of the power of mathematics to unearth nature's secrets. It was possible to conceive of and even to visualize the submarine and the airplane long before technicians produced working models. The notion of a radio wave, on the other hand, would hardly occur even in a flight of fancy and, were it to occur, would be immediately dismissed as such.

Even the man who was most gifted in constructing a physical picture of electromagnetic induction, a picture Maxwell himself used to advance his own thinking, confessed that he was baffled in his attempt to understand the entire phenomenon physically. In a letter to Maxwell written in 1857, Faraday asks whether Maxwell could not express the conclusions of his mathematical work in

> common language as fully, clearly, and definitely as in mathematical formulae? If so, would it not be a great boon to such as I to express them so?—translating them out of their hieroglyphics that we might also work upon them by experiment. . . . If this be possible, would it not be a good if mathematicians, working on these subjects, were to give us the results in this popular, useful, working state, as well as in that which is their own and proper to them?

Unfortunately, Faraday's request cannot be filled to this day.

The inability to explain electromagnetic phenomena qualitatively or materially contrasts sharply with the exact quantitative descriptions furnished by Maxwell and his co-workers. Just as Newton's laws of motion furnished scientists with the means for working with matter and force without explaining either, so Maxwell's equations have enabled scientists to accomplish wonders with electrical phenomena despite a woefully deficient understanding of their physical nature. The quantitative laws are all we have in the way of a unifying, intelligible account. The mathematical formulas are definite and comprehensive; the qualitative interpretation is vague and incomplete. Electrons, electric and magnetic fields, and ether waves merely provide names for the variables that appear in the formulas or, as von Helmholtz stated the point, in Maxwell's theory an electric charge is but the recipient of a symbol.

If physical understanding and the power to reason in physical terms about electromagnetic phenomena are lacking, what is the nature of our grasp of this reality? On what do we base our claim of mastery? Mathematical laws are the only means of probing, revealing, and mastering this large region of the physical world; of such mysterious goings-on mathematical laws are the only knowledge humans possess. Although the answer to these questions is unsatisfactory to the layperson uninitiated into these latter-day Delphic mysteries, the scientist by now has learned to accept it. Indeed, faced with so many natural mysteries, the scientist is only too glad to bury them under a weight of mathematical symbols, bury them so thoroughly that many generations of workers fail to notice the concealment.

We are faced, then, with the amazing fact that one of the largest bodies of scientific theory is almost entirely mathematical. Certain formal deductions from this theory, such as the induction of current in wires or the reception of current hundreds of miles away from a source, can be confirmed by sense impressions, but the body of the theory itself is mathematical.

To some extent we should be prepared for this peculiar state of affairs. After studying Newton's work on gravitation we considered the question: What is gravity and how does it act? We found in that case, too, that we had no physical understanding of the action of gravitation. We have a mathematical law describing the quantitative value of this force and, by using this law and the laws of motion, we can predict effects that can be experimentally checked. The central concept of gravitation, however, remains unknown.

We see, then, that at the heart of our best scientific theories is mathematics or, more accurately, some formulas and their consequences. The firm, bold design of a scientific theory is mathematical. Our mental constructions have outrun our intuitive and sense perceptions. In both theories, gravitation and electromagnetism, we must confess our ignorance of the basic mechanisms and leave the task of representing what we know to mathematics. We may lose pride in making this confession, but we may gain understanding of the true state of affairs. We can appreciate now what Alfred North Whitehead meant when he said, "The paradox is now fully established that the utmost abstractions [of mathematics] are the true weapons with which to control our thought of concrete facts."

The originality of mathematics lies in this paradox since the mathematical sciences establish phenomena that are, apart from human reason, entirely unobvious though physically real. Whitehead has said that cutting out mathematics in human thought is like cutting out not Hamlet but Ophelia. It is true that Ophelia is charming and a little mad, but Hamlet would be more to the point.

In 1931 Einstein described the change in the concept of physical reality after Maxwell as "the most profound and most fruitful that physics has experienced since Newton."

VIII

A Prelude to the Theory of Relativity

> *Common sense is the layer of prejudice laid down in the mind prior to the age of eighteen.*
> Albert Einstein

> *An axiom is a prejudice sanctified by thousands of years.*
> Eric T. Bell

Somewhat as in the case of mathematics proper, the mathematical physicists of about the year 1900 were smug and complacent about their achievements and the state of physical theories. Had they not revealed the existence of a totally new world—the world of electromagnetic phenomena—which was about to enrich, speed up, and extend our cultural and business worlds and to improve human communication? Perhaps the ether that had been accepted for two centuries as the medium through which light and other electromagnetic phenomena were propagated had drugged the mathematical physicists into a restful and uncritical sleep.

However, the complacency of 1900 was the calm before the storm. When the elation brought on by the remarkable achievements had subsided, the mathematical physicists realized that there were major problems still to be resolved. One resolution, the theory of relativity, was to revolutionize the scientific conception of our physical world. As of today this revolution does not have the same impact that radio and television did when they were presented to the public, but for our understanding of the nature of the physical world and of what is objective and real, its implications are as vital.

What problems did the mathematicians and physicists see in the late nineteenth century that sobered them and caused them to undertake a totally new approach to major phenomena of the universe? The first problem was the nature of the geometry of physical space. To understand it we must backtrack somewhat.

During the preceding two thousand years a number of mathematicians questioned the physical truth of Euclid's parallel axiom, which states:

> If a straight line (Figure 32) falling on two straight lines makes the interior angles on the same side less than two right angles, then the two straight lines if extended will meet on that side of the straight line on which the angles are less than two right angles.

That is, if angles 1 and 2 add up to less than 180 degrees, then lines *a* and *b* if produced sufficiently far will meet.

Euclid had good reasons to word his axiom in this manner. He could have asserted instead that, if the sum of the angles 1 and 2 is 180 degrees, then the lines *a* and *b* will never meet; that is, the lines *a* and *b* would be parallel. However, Euclid apparently was afraid to assume that there could be two *infinite* lines that never meet. Certainly neither experience nor self-evidence vouched for the behavior of infinite straight lines. However, Euclid did *prove* the existence of infinite parallel lines on the basis of his parallel axiom and the other Euclidean axioms.

The parallel axiom in the form stated by Euclid was believed to be somewhat too complicated, lacking the simplicity of the other axioms. Apparently even Euclid did not like his version of the parallel axiom, because he did not call on it until he had proved all the theorems he could without it.

Even in Greek times mathematicians began efforts to resolve the problem presented by Euclid's parallel axiom. Two types of attempts were made. The first was to replace the parallel axiom by a seemingly more self-evident statement. The second was to try to deduce it from the other nine axioms of Euclid; were this possible, Euclid's statement

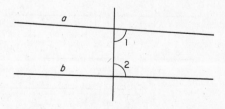

FIGURE 32

would become a theorem and thus no longer be questionable. Over two thousand years dozens of major mathematicians, to say nothing of minor ones, engaged in both types of efforts. The history is long and technical, and most of it will not be retold here because it is readily available and not especially relevant.*

Of the substitute axioms suggested, we should discuss at least one, because it is the one we usually learn in high school today. This version of the parallel axiom, which is credited to John Playfair (1748–1819) and was proposed by him in 1795, states:

> Through a given point P not on a line ℓ (Figure 33), there is one and only one line in the plane of P and ℓ that does not meet ℓ.

All of the other substitute axioms proposed, though seemingly simpler than Euclid's version, were found on closer examination to be no more satisfactory than Euclid's. Of course, Playfair's parallel axiom asserts what Euclid avoided as an axiom: that there can be two *infinite* lines that never meet.

Of the second group of efforts to solve the problem of the parallel axiom, those that sought to deduce Euclid's assertion from the other nine axioms, the most significant attempt was made by Gerolamo Saccheri (1667–1733), a Jesuit priest and professor at the University of Pavia. His thought was that, if one adopted an axiom that differed essentially from Euclid's, then one might arrive at a theorem that contradicted another theorem. Such a contradiction would mean that the axiom denying Euclid's parallel axiom—the only axiom in question—is false, and so the Euclidean parallel axiom must be true—that is, a consequence of the other nine axioms.

In view of Playfair's axiom, which is equivalent to Euclid's, Saccheri assumed first that through the point P (Figure 33) there are no lines parallel to ℓ. From this axiom and the other nine that Euclid adopted, Saccheri did deduce a contradiction. Saccheri tried next the

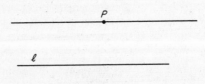

FIGURE 33

*This history can be found in the author's *Mathematics: The Loss of Certainty,* Oxford University Press, 1980.

second and only other possible alternative, namely, that through point
P there are at least two lines p and q that no matter how far produced
do not meet ℓ.

Saccheri proceeded to prove many interesting theorems until he
reached one that seemed so strange and so repugnant that he decided
it was contradictory to the previously established results. Saccheri
therefore felt justified in concluding that Euclid's parallel axiom was
really a consequence of the other nine axioms and so published his
Euclides ab omni naevo vindicatus (Euclid Vindicated from All Faults)
in 1733. However, later mathematicicans realized that Saccheri did
not really obtain a contradiction in the second case, and the problem
of the parallel axiom was still open.

The efforts to find an acceptable substitute for the Euclidean
axiom on parallels or to prove that the Euclidean assertion must be a
consequence of Euclid's nine other axioms were so numerous and so
futile that in 1759 the great mathematician Jean Le Rond d'Alembert
(1717–1783) called the problem of the parallel axiom "the scandal of
the elements of geometry."

Gradually the mathematicians began to approach the correct
understanding of the status of Euclid's parallel axiom. In his doctoral
dissertation of 1763 Georg S. Klügel (1739–1812), later a professor at
the University of Helmstädt, made the remarkable observation that
the certainty with which people accepted the truth of the Euclidean
parallel axiom was based on experience. This observation introduced
for the first time the thought that experience rather than self-evidence
substantiated the axiom. Klügel expressed doubt that the Euclidean
assertion could be proved. Moreover, he realized that Saccheri had not
arrived at a contradiction but merely at results that were strange.

Klügel's paper suggested further work on the parallel axiom to
Johann Heinrich Lambert (1728–1777). In his book, *Theorie der Par-
allellinien,* written in 1766 and published in 1786, Lambert, somewhat
like Saccheri, considered the two alternative possibilities. He, too,
found that by assuming no lines through P parallel to ℓ (Figure 33) we
obtain a contradiction. However, unlike Saccheri, Lambert did not
conclude that he had obtained a contradiction from assuming at least
two parallels through P. Moreover, he realized that any body of
hypotheses that did not lead to contradictions offered a possible geometry. Such a geometry would be a valid logical structure even though
it might have little to do with physical figures.

The work of Lambert and other men such as Abraham G. Kästner
(1719–1800), a professor at Göttingen who was Gauss's teacher, war-

rants emphasis. They were convinced that Euclid's parallel axiom could not be proven on the basis of the other nine Euclidean axioms; that is, it is independent of Euclid's other axioms. All three recognized the possibility of a non-Euclidean geometry—that is, one whose axiom on parallel lines differed essentially from Euclid's.

The most distinguished mathematician who worked on the problem posed by Euclid's parallel axiom was Karl Friedrich Gauss (1777–1855). Gauss was fully aware of the vain efforts to establish Euclid's parallel axiom, because this was common knowledge in Göttingen. However, even up to 1799 Gauss still tried to deduce Euclid's parallel axiom from other more plausible assumptions, and he still believed Euclidean geometry to be the geometry of physical space, although he could conceive of other logical non-Euclidean geometries. But on December 17, 1799 Gauss wrote to his friend and fellow mathematician Wolfgang Bolyai (1775–1856):

> As for me I have already made some progress in my work. However, the path I have chosen does not lead at all to the goal which we seek [deduction of the parallel axiom], and which you assure me you have reached. It seems rather to compel me to doubt the truth of geometry itself. It is true that I have come upon much which by most people would be held to constitute a proof [of the deduction of the Euclidean parallel axiom from the other axioms]; but in my eyes it proves as good as nothing.

From about 1813 on, Gauss developed his non-Euclidean geometry, which he first called anti-Euclidean geometry, then astral geometry, and finally non-Euclidean geometry. He became convinced that it was logically consistent, and he was rather sure that it might be applicable to the physical world.

In a letter to his friend Franz Adolph Taurinus (1794–1874) of November 8, 1824 Gauss wrote:

> The assumption that the angle sum [of a triangle] is less than 180 degrees leads to a curious geometry, quite different from ours [Euclidean] but thoroughly consistent, which I have developed to my entire satisfaction. The theorems of this geometry appear to be paradoxical, and, to the uninitiated, absurd, but calm, steady reflection reveals that they contain nothing at all impossible.

We shall not discuss the specific non-Euclidean geometry that Gauss created. He started to, but did not write up a full deductive presentation, and the theorems he did prove are much like those we shall encounter in the work of Lobatchevsky and Bolyai. He said in his letter of January 27, 1829 to the mathematician and astronomer Friedrich Wilhelm Bessel (1784–1846) that he probably would never publish his findings on this subject because he feared ridicule or, as he put it, he feared the clamor of the Boeotians, a figurative reference to a dull-witted Greek tribe. One must remember that although a few mathematicians had been gradually reaching the denouement of the work in non-Euclidean geometry, the intellectual world at large was still dominated by the conviction that Euclidean geometry was the only possible geometry. What we do know about Gauss's work in non-Euclidean geometry is gleaned from his letters to friends, two short reviews in the *Göttingische gelehrte Anzeigen* of 1816 and 1822, and some notes of 1831 found among his papers after his death.

The two men who receive more credit for the creation of non-Euclidean geometry are Lobatchevsky and Bolyai. Actually their work was an epilogue to prior innovative ideas, but because they published systematic deductive works they are usually hailed as the creators of non-Euclidean geometry. Nikolai Ivanovich Lobatchevsky (1793–1856), a Russian, studied at the University of Kazan and from 1827 to 1846 was professor and rector at that university. From 1825 on he presented his views on the foundations of geometry in many papers and two books. Johann Bolyai (1802–1860), son of Wolfgang Bolyai, was a Hungarian army officer. He published a twenty-six-page paper on non-Euclidean geometry, "The Science of Absolute Space," which he called absolute geometry, as an appendix to the first volume of his father's two-volume book *Tentamen juventutem studiosam in elementa Matheseos.* Although this book appeared in 1832–1833 and thus followed publications by Lobatchevsky, Bolyai seems to have worked out his ideas on non-Euclidean geometry by 1825 and was convinced by that time that the new geometry was not self-contradictory.

Gauss, Lobatchevsky, and Bolyai had realized that the Euclidean parallel axiom could not be proven on the basis of the other nine axioms and that some additional axiom was needed to found Euclidean geometry. Because the parallel axiom was an independent fact, it was then at least logically possible to adopt a contradictory statement and develop the consequences of the new set of axioms.

The technical content of what these men created is rather simple. It is just as well to note Lobatchevsky's work, because all three did about the same thing. Lobatchevsky boldly rejects Euclid's parallel axiom and makes the assumption that, in effect, Saccheri had made. Given a line *AB* and a point *P* (Figure 34), then all lines through *P* fall into two classes with respect to *AB,* namely, the class of lines that meet

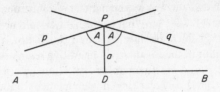

FIGURE 34

AB and the class of lines that do not. More precisely, if *P* is a point at a perpendicular distance *a* from the line *AB*, then there exists an acute angle *A* such that all lines that make with the perpendicular *PD* an angle less than *A* will intersect line *AB*, while lines that make an angle greater than or equal to *A* do not interesct line *AB*. The two lines *p* and *q* that make the angle *A* with line *AB* are the parallels, and *A* is called the angle of parallelism. Lines through *P* other than the parallel lines that do not meet line *AB* are called nonintersecting lines, although in Euclid's sense they are parallel lines. In this sense, Lobatchevsky's geometry contains an infinite number of parallels through *P*.

He then proved several key theorems. If angle *A* equals $\pi/2$, then the Euclidean parallel axiom results. If angle *A* is acute, then it follows that angle *A* increases and approaches $\pi/2$ as the perpendicular *a* decreases to zero; furthermore, angle *A* decreases and approaches zero as *a* becomes infinite. The sum of the angles of a triangle is always less than 180 degrees and approaches 180 degrees as the area of the triangle decreases. Moreover, two similar triangles are necessarily congruent.

The most significant fact about this non-Euclidean geometry is that *it can be used to describe the properties of physical space as accurately as Euclidean geometry does.* Euclidean geometry is not the necessary geometry of physical space; its physical truth cannot be guaranteed on any a priori grounds. This realization, which did not call for any technical mathematical development, because this had already been done, was first attained by Gauss.

But no one yields up treasures readily. Gauss apparently reconsidered the matter of the truth of mathematics and saw the rock to which he could anchor it. In a letter to Heinrich W. M. Olbers (1758–1840), written in 1817, he said,

> I am becoming more and more convinced that the physical necessity of our Euclidean geometry cannot be proved, at least not by human reason nor for human reason. Perhaps in another life we shall be able to obtain

insight into the nature of space which is now unattainable. Until then we must not place geometry in the same class with arithmetic, which is purely a priori, but with mechanics.

Gauss, unlike Kant, did not accept the laws of mechanics as truths. Instead, he and most others followed Galileo in believing that these laws are founded on experience. Gauss was asserting that truth lies in arithmetic and consequently in algebra and analysis, which are built on arithmetic, because the truths of arithmetic are clear to our minds.

Lobatchevsky also considered the applicability of his geometry to physical space and did give an argument to show that it could be applicable to very large geometrical figures. Thus, by the 1830s, not only was a non-Euclidean geometry recognized, but its applicability to physical space was also considered possible.

For thirty years or so after the publication of Lobatchevsky's and Bolyai's works, the mathematicians ignored this non-Euclidean geometry, regarding it as a logical curiosity. Some mathematicians did not deny its logical coherence. Others believed that it must contain contradictions and so was worthless. Almost all mathematicians maintained that the geometry of physical space, *the* geometry, must be Euclidean. William R. Hamilton (1805–1865), certainly one of the outstanding mathematicians of the time, expressed his objection to non-Euclidean geometry in 1837:

> No candid and intelligent person can doubt the truth of the chief properties of *Parallel Lines,* as set forth by Euclid in his *Elements,* two thousand years ago; though he may well desire to see them treated in a clearer and better method. The doctrine involves no obscurity nor confusion of thought, and leaves in the mind no reasonable ground for doubt, although ingenuity may usefully be exercised in improving the plan of the argument.

In his presidential address of 1883 to the British Association for the Advancement of Science, Arthur Cayley (1821–1895) affirmed Hamilton's view:

> My own view is that Euclid's twelfth axiom [usually called the fifth or parallel axiom] in Playfair's form of it does not need demonstration, but is part of our notion of space, of the physical space of our experience—which one becomes acquainted with by experience, but which is the representation lying at the foundation of all external experience. . . . Not that the propositions of geometry are only approximately true, but that they remain absolutely true in regard to that Euclidean space which has been so long regarded as being the physical space of our experience.

Felix Klein (1849–1925), one of the truly great mathematicians of recent times, expressed about the same view. Although Cayley and Klein had themselves worked in non-Euclidean geometries (of which there are several, as we shall see), they regarded them as novelties that result when artificial new distance functions are introduced in Euclidean geometry. They refused to grant that a non-Euclidean geometry is as basic and as applicable as Euclidean. Of course, their position, in pre-relativity days, was tenable.

Unfortunately, the mathematicians had abandoned God, and so the Divine Geometer refused to reveal which geometry He had used to design the universe. The mathematicians were thrown on their own resources. The mere fact that there can be alternative geometries was in itself a shock, but the greater shock was that one could no longer be sure of whether a non-Euclidean geometry might prove to be applicable to physical space.

The problem of choosing a geometry that fits physical space, originally raised by the work of Gauss, stimulated another creation that gave the mathematical world still further inducement to believe that the geometry of physical space could be non-Euclidean. The creator was Georg Bernhard Riemann (1826–1866), a student of Gauss and later professor of mathematics at Göttingen. Although the details of Lobatchevsky's and Bolyai's work were unknown to Riemann, they were known to Gauss, and Riemann certainly knew Gauss's doubts about the truth and necessary applicability of Euclidean geometry.

Gauss paved the way to Riemann's sensational ideas by introducing another revolutionary idea. Normally we study the geometry on the surface of a sphere as part of three-dimensional Euclidean geometry, and thereby no radical ideas are introduced. But suppose one were to consider the surface of a sphere as a space unto itself and build a geometry that would fit that space. Rectangular coordinates could not be used, because these call for straight lines, which do not exist on a sphere. One might hit on the idea of using latitude and longitude as coordinates of the points. The next concern might be to determine the paths of shortest length from one point to another. Experience interpreted by the all-wise mathematicians soon led to the conclusion that the shortest paths are arcs of great circles such as the circles of longitude or, in fact, any circle whose center is at the center of the Earth. These circles would be the "lines" of the geometry. Continuing to investigate the geometry of this spherical surface, one would find many strange theorems. For example, a triangle formed by the arcs of great circles, the "line" segments of the geometry, would have an angle sum of more than 180 degrees.

What Gauss was suggesting in his famous paper of 1827 was that, if we study surfaces as independent spaces, then the two-dimensional geometries that would fit those surfaces would indeed be strange and would depend on the shapes of these surfaces. Thus the ellipsoidal surface, which roughly has the shape of a football, would have a geometry quite different from that of the sphere.

What about parallel "lines"? Obviously, because any two great circles meet not only once but twice, we should have to have as an axiom that any two "lines" meet in two points. It became quite clear that the geometry of the surface of a sphere would be what was later recognized as a new non-Euclidean geometry called *double elliptic*. This would be the natural geometry for the surface of the Earth and, indeed, it is as practical and at least as convenient as treating the sphere as a surface in three-dimensional Euclidean geometry.

The ideas of Gauss were familiar to Riemann. Gauss had assigned several possible subjects to Riemann for the lecture he was required to deliver to qualify for the title of Privatdozent, a teaching position at the University of Göttingen. Riemann chose the foundations of geometry, and he gave this lecture in 1854 to the philosophical faculty at Göttingen with Gauss present. It was published in 1868 under the title "On the Hypotheses Which Lie at the Foundation of Geometry."

The investigation of the geometry of physical space offered by Riemann reconsidered the entire problem of the structure of space. Riemann first took up the question of just what is certain about physical space. What conditions or facts are presupposed in the very concept of space before we determine by *experience* the particular axioms that hold in physical space? From these conditions or facts, treated as axioms, he planned to deduce further properties. These axioms and their logical consequences would be a priori and necessarily true. Any other properties of space would then have to be learned empirically. One of Riemann's objectives was to show that Euclid's axioms were indeed empirical rather than self-evident truths. He adopted the analytical approach (algebra and the calculus), because in geometrical proofs we may be misled by our perceptions to assume facts not explicitly recognized as assumptions.

Riemann's search for what is a priori led him to study how a space behaves locally, because its properties may vary from point to point. This approach is called *differential geometric* as opposed to the consideration of space as a whole, as one finds it in Euclid or in the non-Euclideam geometry of Gauss, Bolyai, and Lobatchevsky.

Riemann's approach to geometry led him to define distance between two typical or generic points, whose corresponding coordinates differ only by infinitesimal amounts. This distance he denoted by ds. He assumed that the square of this distance in three dimensions (although he considered n dimensions) was:

$$ds^2 = g_{11}dx_1^2 + g_{12}dx_1dx_2 + g_{13}dx_1dx_3 + g_{21}dx_1dx_2 + g_{22}dx_2^2 \\ + g_{23}dx_2dx_3 + g_{13}dx_1dx_3 + g_{23}dx_2dx_3 + g_{33}dx_3^2$$

wherein the g_{ij} are functions of the coordinates x_1, x_2, and x_3, $g_{ij} = g_{ji}$, and the right side is always positive for all possible values of the g_{ij}. This expression for ds is a generalization of the Euclidean formula:

$$ds^2 = dx_1^2 + dx_2^2 + dx_3^2$$

which in itself is the differential form of the Pythagorean theorem. By allowing the g_{ij} to be functions of the coordinates, Riemann provided for the possibility that the nature of the space may vary from point to point. From this formula for ds^2, one can deduce many facts about length, area, volume, and other quantities by the methods of the calculus.

In this same address, Riemann made many more significant points. He adds, "It remains to resolve the question of knowing in what measure and up to what point these a priori hypotheses about space are confirmed by experience." The properties of *physical* space are to be obtained only from experience. In particular, the axioms of Euclidean geometry may be only approximately true of physical space. He ends his paper with the prophetic remark:

> Either therefore the reality which underlies space must form a discrete manifold or we must seek the ground of its metric relations outside it, in the binding forces which act on it. This leads us into the domain of science, that of physics, into which the object of our work does not allow us to go today.

Here Riemann suggested that the nature of real space must take into account physical phenomena that take place in space. This profound thought might have been elaborated if Riemann had not died at the age of forty.

The point was developed somewhat by the mathematician William Kingdon Clifford (1854–1879). Some physical phenomena, Clifford believed, result from variations in the curvature of space. The curvature changes not only from place to place but from time to time, the latter being a result of the motion of matter. Space is analogous to a hilly surface, and the ordinary laws of Euclidean geometry are not valid in such a space. A more exact investigation of physical laws would not be able to ignore these "hills" in space.

Clifford wrote in 1870:

> I hold in fact: (1) That small portions of space are of a nature analogous to little hills on a surface which is on the average flat. (2) That this property of being curved or distorted is continually passed on from one portion of space to another after the manner of a wave. (3) That this variation of the curvature of space is really what happens in that phenomenon which we call the motion of matter, whether ponderable or ethereal. (4) That in this physical world nothing else takes place but this variation, subject, possibly, to the law of continuity.

Clifford also suggested the possibility that the effects of gravitation might be due to the curvature of space, but spatial measurements at that time did not confirm his suggestion. Indeed, his idea, brilliant as it was, had to await the work of Einstein.

What Riemann and Clifford were suggesting can be made more understandable if we consider the natural geometry of the surface of the Earth in a mountainous region. On the surface of such a region there may be no straight lines. Moreover, whatever curve does give the shortest distance between two points is almost always not a straight line. Furthermore, these shortest paths, or geodesics, need not have the same shape. These mountain dwellers might then proceed to consider triangles. That is, given three points and arcs of geodesics connecting these points, what properties would these triangles possess? Obviously, the properties would depend on the shape of the terrain encompassed by the geodesic arcs. Some triangles might contain angle sums much larger than 180 degrees, others much less than 180 degrees. There is no doubt that these people would arrive at a non-Euclidean geometry. One important characteristic of the geometry is that it would be non-homogenous; that is, the properties of the figures of this geometry would vary from place to place, just as the shape of the mountainous surface itself varies from place to place.

The material in Gauss's notes became available after his death in 1855 when his reputation was unexcelled, and the publication in 1868 of Riemann's 1854 paper convinced some mathematicians that a non-

Euclidean geometry could be the geometry of physical space and that we could no longer be sure of which geometry was the correct one.

Non-Euclidean geometry and its implication about the physical truth of geometry were accepted gradually by mathematicians, but not because the arguments for its applicability were strengthened in any way. Instead, the reason was given in the early 1900s by Max Planck, the founder of quantum mechanics: "A new scientific truth does not triumph by convincing its opponents and making them see light, but rather because its opponents eventually die, and a new generation grows up that is familiar with it."

We have been discussing how mathematicians began to be troubled about the geometry of physical space. Another problem began to trouble the late nineteenth-century mathematical physicists. One of the most firmly embedded assumptions in the scientific thinking of the eighteenth and nineteenth centuries was that a force of gravity exists. According to Newton's first law of motion, a body at rest remains at rest and a body in motion continues to move at a constant velocity along a straight line unless acted on by a force. Hence, were there no force of gravity, a ball held in the hand and merely released would remain suspended in the air. Similarly, were there no force of gravity, the planets would shoot out into space along straight-line paths. Such strange phenomena do not occur. The universe acts *as if* there were a force of gravity.

Although Newton did show that the same quantitative law covered all the terrestrial and celestial effects of gravity's action, the physical nature of the force of gravity had never been understood. How does the sun, 93 million miles away from the Earth, exert its pull on the Earth, and how does the Earth exert its pull on the variety of objects near its surface? Although there were no answers to these questions, the physicists were not perturbed. Gravity was such a useful concept that they were content to accept it as a physically real force. Indeed, if it were not for other more pressing questions and difficulties that arose around 1880, the complacency of physicists on the subject of gravity might not yet have been seriously ruffled.

Another problem raised by the introduction of the force of gravity had also been quietly thrust aside. Every physical object possesses two apparently distinct properties, mass and weight. Mass is the resistance an object offers to a change in its speed or direction of motion. Weight is the force with which the Earth attracts an object. Under the Newtonian theory the mass of an object is constant, whereas its weight depends on how far the object is from the center of the Earth. At the center the mass is the same but the weight is zero. On the surface of the moon the mass is still the same but the gravitational pull of the moon is one-eightieth that of the Earth. However, the distance from the center of attraction is only one-quarter of that for the Earth. In view of the inverse-square law of distance in the law of gravitation (Chapter VI), the gravitational pull of the moon on an object on its surface is sixteen times stronger. The result of both effects is that the weight of an object on the moon is $1/80 \times 16$, or one-fifth of the weight on the surface of the Earth. Astronauts in a spaceship have the same mass as they do on Earth, but in the spaceship they have no weight.

Although these two properties of matter, mass and weight, are distinct, the ratio of the two is always the same at a given place. This fact is as surprising as if the ratio of coal production to wheat production were exactly the same each year. Were coal and wheat production actually related in this way, we should look for an explanation in the economic structure of the nation. In like manner, an explanation of the constant ratio of weight to mass was called for. Until Einstein's day, none had been found.

One more physical assumption must be mentioned before we examine Einstein's work. Attempts to explain the nature of light date back to Greek times. In particular, since the early nineteenth century the most commonly accepted view of light regarded it as a wave motion much as sound is. Because it is not possible to conceive of a wave motion without a medium to carry the wave, scientists reasoned that there must be a medium that carries light. However, there was no evidence that the space through which light travels, whether from the distant stars or from the sun, contained any material substance to transmit the waves. Therefore, scientists assumed the existence of a new "substance," *ether,* which could be neither seen, tasted, smelled, weighed, nor touched. Moreover, ether had to be a fixed medium existing throughout all space through which the Earth and other heavenly bodies moved freely as in a vacuum. Thus, the properties ether was supposed to possess were contradictory (see Chapter VII).

Despite the many dubious and poorly understood assumptions that lay at the foundation of late nineteenth-century physics, no group of scientists in any age was ever more cocksure that it had discovered the laws of the universe. The eighteenth century had been optimistic; the nineteenth was supremely confident. Two hundred years of partial success had so turned the heads of the scientists and philosophers that Newton's laws of motion and the law of gravitational attraction were declared to be immediate consequences of the laws of thought and

pure reason. The word *assumption* did not appear in scientific literature even though, as Newton had expressly stated, the concepts of
gravitation and ether were hypotheses and, indeed, hypotheses not at
all understood physically. Nevertheless, what was inconceivable to
Newton was, to the late nineteenth century, inconceivable otherwise.

The Relativistic World

> The great architect of the Universe now begins to appear as a
> pure mathematician.   James H. Jeans

> The general laws of nature are to be expressed by equations
> which hold good for all systems of coordinates.
>
> Albert Einstein

The drastic overhauling of physics began inauspiciously when American physicists decided in 1881 to check that the Earth moves through a stationary ether. Albert A. Michelson (1852–1931) devised an experiment based on a very simple principle.

A little arithmetic shows that it takes longer to row a given distance down a river and then back if there is a current than if there is no current. (We touched on this concept in Chapter I in the discussion of intuition.) For example, if a man can row at the rate of four miles per hour in still water, then, with no current present, he can go twelve miles down and then twelve miles back in six hours. If a current flows at the rate of two miles per hour, however, the man's progress downstream will be at the rate of four plus two miles per hour, while his rate upstream will be four minus two miles per hour. At these rates his total time for the trip will be two plus six, or eight hours. The principle involved here is that if a constant velocity, such as the velocity of the stream, hinders a motion for a longer time than it helps the motion, the net result is a loss in time.

Michelson and a later collaborator, Edward W. Morley (1838–1923), used this principle in the following way. From a point *A* (Figure 35) on the Earth, a ray of light was sent to a mirror placed on the Earth at *B;* the direction from *A* to *B* was the direction of the Earth's motion around the sun. The ray was expected to travel through the ether to *B*

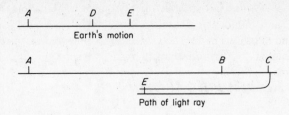

FIGURE 35

at the usual velocity with which light travels and then be reflected back toward *A*. Because of the Earth's motion, however, the mirror at *B* moves to a new position *C* while the light ray is traveling toward it. Hence the Earth's motion delays the light ray in reaching the mirror. At *C* the ray is reflected back toward *A*. While the ray is traveling *toward* point *B*, however, the Earth carries point *A* to a new position *D*, and while the ray is traveling *back*, the Earth carries point *D* to yet another position *E*. Therefore, the motion of the Earth helps the light ray in going from *C* to *E*. However, the distance traveled from *C* to *E* is shorter than the distance from *A* to *C*. In this way, the light ray is helped by the Earth's motion for a shorter time than it is delayed on the way out. The Earth's motion has the same effect as the velocity of the stream in the previous example. Hence, by the principle described there, the light ray should require *more* time to make the journey from *A* to *C* to *E* than if it had traveled twice the distance *AB* with the Earth stationary in the ether. Although they used a very ingenious and delicate testing device known as an interferometer, Michelson and Morley were unable to detect the increase in time. The motion of the Earth through the ether was apparently not taking place.

Physicists were faced with an inescapable dilemma. The ether that was needed to carry light had to be a fixed medium through which the Earth moved; yet this supposition was inconsistent with the result of experimentation. The failure of theory to agree with such a fundamental experiment could not be ignored. By this time physicists were convinced that some of their science needed overhauling.

One more related difficulty faced the nineteenth-century mathematical physicists. To understand this let us digress for a moment. Newton believed in absolute space and time and defined them thus in his *Mathematical Principles of Natural Philosophy:* "Absolute space, in its own nature and without regard to anything external, remains similar and immovable. Absolute, true and mathematical time, of itself and of its own nature, flows uniformly on without regard to anything external." These concepts he regarded as having objective reality, apart from material bodies or human experiences, and they are known, he believed, to a superhuman observer, God. Furthermore, the ideal formulation of the mathematical and scientific laws of this universe are the laws God can obtain by His absolute measurements. It was only by knowing the motion of the Earth relative to the fixed observer, God, that humans could translate His laws into the true form. We see, then, that Newtonian scientific thought was based fundamentally on metaphysical assumptions involving God, absolute space, absolute time, and absolute laws. Many of Newton's contemporaries and successors, notably Euler and Kant, believed in these concepts.

Of course, Newton realized that humans do not have knowledge of absolute space and absolute time. He therefore presupposed that there are inertial observers, those for whom Newton's first law of motion is valid. This law, we may recall, states that if no force acts on an object, the object if at rest will remain at rest or, if moving, moves with a constant velocity in a straight line.

Given any one inertial observer, many others might be found who were at rest or moving, with respect to each other, with a constant velocity and in a straight line. All of these observers moved in what are called inertial frames of reference. Let us consider this concept using a simple example. Suppose a passenger on a ship moving with constant velocity moves from one position to another with a constant speed and measures the distance he has moved. Now suppose someone on shore measures the distance from the passenger's initial position to his final one. Of course, relative to the shore the distance is greater. If the motion of the ship is taken into account, the discrepancy can be explained. Clearly, there are two frames of reference; one is that of the person on shore, and the other is that of the passenger on board.

Consider two such frames moving relatively to one another with a uniform translational motion, and suppose that an object is moving with respect to both frames. With respect to the first frame, the object describes a particular path with some definite motion along that path. With respect to the second frame, the path and motion will be different. For mathematical purposes, one uses a coordinate system to specify the desired frame of reference. In Figure 36 we have supposed that the K frame is fixed and the K' frame is moving to the right with a

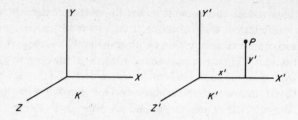

FIGURE 36

constant velocity relative to K. Both observers are supposed to have identical clocks.

Now let P be a point in space. Its coordinates with respect to K' are x' and y'; with respect to K they are x and y. Because the right-hand frame moves with velocity v, the relation between x and x' is given by $x = x' + vt$. This equation transforms the abscissa of P relative to K' to the abscissa of P relative to K. Moreover, $y = y'$. Also, if both observers measure time intervals in the same way, then

$$t = t'$$

In Newtonian physics all mechanical laws are unaltered by such a transformation; that is, if expressed in x, y, t coordinates, the law is the same in x', y', t' coordinates provided the velocity of the second system relative to the first is constant.

The two frames are called Galilean or inertial frames. One moves with respect to the other at a constant speed. No relative acceleration or rotation between these frames is to enter. In Newton's terms, Galilean frames are at rest or moving with uniform translational speed through absolute space but without acceleration or rotation. Which one is at rest in absolute space cannot be determined, but because we know the laws of transformation this does not matter. Moreover, the differential equations that hold in one frame also hold in the other. To repeat, the classical laws of mechanics are the same in both.

Next let us consider Maxwell's equations. At the close of the nineteenth century it was believed that the same partial differential equations held in any Galilean frame. Thus, the same situation seemed to obtain in electromagnetics as in Newtonian mechanics. However, this belief led to a contradiction. In order to obtain the laws holding in frame K' that hold in frame K, we apply the laws of transformation to

K. If this is done for the equations of electromagnetics, we find that we must modify them by adjoining terms that involve the relative velocity of the two frames. The reason is simply that velocity is not invariant and Maxwell's equations involve c, the velocity of light. For example, consider a light signal propagating to the right with velocity c and another propagating to the left with velocity c. An observer moving to the right with velocity v is "catching up" to the light signal, and so for that observer the signal has the velocity $c - v$. On the other hand, this observer is running away from the second light signal, and so his velocity relative to this signal is $c + v$. For the moving observer the two light signals do not propagate with the same velocity, and so Maxwell's equations do not have the same form for him. As far as Maxwell's equations are concerned, there is only one preferred frame: the frame at rest with respect to the ether.

Thus, the transformation of Maxwell's equations from one frame of reference to another moving with a constant velocity with respect to the first frame showed that Maxwell's equations did not behave in the same way as Newton's laws of mechanics. In the latter a simple transformation converts from one system to another, but this is not so for Maxwell's equations.

The distinguished mathematical physicist Hendrick Antoon Lorentz (1853–1928) thought of a possible solution. What if we retain the invariance of Maxwell's equations and modified the law of transformation from one reference frame to another. For simplicity we assume that only one space dimension and time are altered. Lorentz obtained the following equations for the transformation from one coordinate system to another moving with constant velocity v with respect to the first one:

$$x' = \frac{x - vt}{\sqrt{1 - v^2/c^2}}, \ y' = y, \ z' = z, \ t' = \frac{t - vx/c^2}{\sqrt{1 - v^2/c^2}}.$$

These equations assume that the second reference frame is moving in the same direction as the first one, namely, the x direction. We note that in the Lorentz equations distance and time are involved together. Moreover, the relations between x and x' and between t and t' are not the simple ones in the Galilean transformation. In particular, t and t' are not simultaneous, that is, t does not equal t'. Let us note also that c is the velocity of light, which is 186,000 miles per second; the velocities v that humans normally encounter are relatively so small that the Lorentz equations practically reduce to the Galilean transformation.

In 1905 Albert Einstein (1879–1955) entered the picture. Einstein leaned toward physics as opposed to mathematics. Although he knew a lot of mathematics and learned more as he went along, it was just a tool for him. Physics was more important. He was impressed with the work in electromagnetic theory and by Heinrich Hertz's work in particular. Despite the revolutionary character of his twentieth-century work in relativity and, as we shall note in the next chapter, in quantum mechanics, he was the last of the great nineteenth-century thinkers who invoked mathematics as no more than an aid to physical thinking. Truth to him transcended mathematics. Nevertheless, his theory of relativity rests entirely on mathematics.

Having studied Lorentz's work and the Michelson–Morley experiment (although there is some question about how much he knew of both), Einstein sought to eliminate the apparent disagreement between classical mechanics and electromagnetic theory and to resolve some of the other problems we have mentioned (see Chapter VIII). One of his 1905 papers was entitled "On the Electrodynamics of Moving Bodies." This paper contains what is now known as the special theory of relativity. Basically, one can say that the more limited special theory of relativity was born in electromagnetic theory.

Einstein took the bull by the horns when he made several assumptions. Because there was no way of determining absolute space and time but only inertial frames, he assumed that in mechanics, too, the transformation from one inertial frame to another should not be Newton's but Lorentz's. This decision was not willful or arbitrary. Lorentz had sought to secure the invariance of Maxwell's equations under coordinate transformations. Einstein believed that he could extend the range of Newton's laws, even if just for inertial frames. The fact that the speed of light seemed to be the same for all observers (regardless of the motion of the light's source) also influenced him, and this also became a postulate in his special theory of relativity. Moreover, because electromagnetic fields exert force on electrons, and force is a mechanical concept, there was good reason to believe that the Lorentz equations should apply to mechanics. The concept of ether he abandoned. Just how light is transmitted was and remains open. Goethe once wrote that the greatest art in theoretical and practical life consists in changing a problem into a postulate. This is what Einstein did in 1905.

Let us now consider some deductions from Einstein's postulates of his special theory of relativity. The first is that two observers, one moving in a straight line with constant velocity v with respect to the other, will not agree on the simultaneity of events. Let us consider a somewhat mundane example.

Suppose that a passenger in the middle of a long, fast-moving train *simultaneously* sees two flashes of light, one of which emanates from a spot in the front car of the train and the other from the car at the rear. An observer standing alongside the track halfway between the front and rear of the train also sees the two flashes but *not simultaneously*. The one from the rear reaches the observer first. The question for consideration is: Were the flashes *emitted* simultaneously?

Both observers would agree that they were not. Because the observer on the ground is positioned exactly between the flashes, the two light rays must travel the same distance and, therefore, take the same amount of time to reach him. Because this observer saw the flash from the rear first, this flash must have occurred first. The passenger on the train would reason that from his viewpoint the velocity of the light ray coming from the rear is the velocity of light minus the velocity of the train. On the other hand, relative to the passenger the velocity of the ray from the front is the velocity of light plus the velocity of the train. Because both rays traveled half the train's length to reach him, and because the ray from the rear required more time, the flash from the rear must have been sent out first to allow the two rays to reach him simultaneously. There seems to be no difficulty whatever in this situation.

The two observers agreed on the order of the flashes, because they both assumed that the man on the ground was at rest with respect to the ether while the passenger on the train was in motion. Suppose, however, that the passenger on the train should take the unorthodox view that the train is at rest with respect to the ether and that the Earth is moving toward the rear of the train. According to this view the passenger on the train would correctly conclude that because he saw the flashes simultaneously, they were emitted simultaneously. The observer on the ground would undoubtedly prefer to stand by his previous position, namely, that he and the Earth are at rest with respect to the ether and that the flash from the rear car occurred first. We now have disagreement on the simultaneity of the two flashes arising out of disagreement about who is at rest with respect to the ether. Who is?

Unfortunately, the passenger on the train is as much entitled to the belief that the train is at rest with respect to the ether as the observer on the ground is that the Earth is stationary in ether, for the Michelson–Morley experiment shows us that we cannot detect any motion through the ether. It follows that *two observers moving relative to each other must disagree on the simultaneity of two events.*

If two observers disagree about the simultaneity of two events, they must also disagree on the measure of distances. Suppose an observer on Mars and one on Earth agree to measure the distance from Earth to the sun. Because this distance is variable they must agree to measure it at a given instant. For both observers to agree on the given instant, however, both must agree on the simultaneity of occurrences, such as the striking of clocks, which mark the instant. Because two observers moving relative to each other will not agree on the simultaneity of these occurrences, they will obtain different measures of the distance from Earth to the sun "at a given instant."

Even the nature of the path pursued by an object will depend on the observer. Let us consider another simple example. A stone dropped from a train moving with constant velocity will seem to a passenger on the train to fall in a straight line, but to an observer on the ground the stone will seem to move along a parabolic path. In other words, the trajectory varies with the observer.

Two observers moving with respect to each other will disagree not only on the measure of distances but also on the measure of time intervals. Otherwise, the observers would have to agree on the simultaneity of events that mark the beginning of the interval as well as on those marking the end; this they cannot do.

Einstein made further deductions. If one observer is stationary and another is moving with respect to this observer with constant velocity v in a fixed direction—if, for example, the moving observer is on a train—then length on the moving body is judged by the stationary observer to be shorter, and vice versa. As for time, the stationary observer finds that the observer moving with respect to the Earth, say, moves more slowly. The moving man's cigar seems to the stationary observer to last twice as long as his own. Put otherwise, a clock in a frame S' is at rest in S'. Seen from another frame S, the clock in S' runs slow by $(1 - 1/\beta)$ per second, where $\beta = \sqrt{1 - v^2/c^2}$. The converse is also true. In general, the relationship between the two frames is the Lorentz transformation. Moreover, we cannot separate the measurement of space and time except for any one observer, just as we cannot separate horizontal and vertical for all observers.

It should be emphasized that we are not speaking of the effect of distance on sight or of optical illusions when we discuss the differences that different observers might obtain in the measurement of lengths.

Nor are we speaking of a psychological or emotional effect when we speak of disagreement on time intervals.

To consider a numerical example, an observer on the Earth would find that a rocketship moving at the speed of 161,000 miles per second relative to the Earth is half as large as a person on the ship would find it to be. And a clock on such a rocket would be judged to "move half as fast" by the earthbound observer as it would for the person in the rocket. An observer in the rocket would draw the same conclusions on size and time for objects and events on the Earth. Furthermore, both sets of measurements are correct, each in its own space and time world.

We have in this doctrine of local length and local time one of the startlingly new assertions of the theory of relativity. The strangenesss of the ideas should not blind us to the fact that they agree far more with experiment and the reasoning on simultaneity, which we examined above, than do the absolute notions of Newton. Indeed, if they did not, scientists would not hold them for a moment, relative or absolute. These relations of length and time as seen by one observer relative to an observer moving with velocity v can be deduced from the Lorentz transformation.

Another consequence of the postulates of special relativity concerns the addition of velocities. Suppose a man can row in still water at four miles per hour and the current flows at two miles per hour. Is the resulting velocity six miles per hour? Not according to special relativity. The resultant velocity V is, in general terms:

$$V = \frac{u + v}{1 + uv/c^2}$$

where u and v are the two separate velocities. An interesting feature of this formula is that if $u = c$, then $V = c$.

Perhaps the strangest implication of the special theory of relativity is that the mass of any object increases with its velocity. This subject Einstein treated in a fourth paper of 1905. The mathematical statement, if we denote by m the mass at rest with respect to an observer, is

$$M = \frac{m}{\sqrt{1 - v^2/c^2}} \qquad (1)$$

where M is the mass of the moving object and v is its velocity. Now how can this be? Surely when the velocity of the mass increases, it does not add molecules. The answer is a surprising one. To a good approximation one can show that the *increase* in mass is very nearly the kinetic energy of the rest mass divided by c^2. Roughly speaking, the addition to the mass amounts to energy. One can say that the moving mass behaves as though its mass has increased, but the increase is physically an amount of energy.

The above relation of mass to energy seems incredible, but it is part of our almost everyday experience. Let us consider first the conversion of mass to energy. We all have occasion to use flashlights. Here we convert the mass in the batteries to light, which has energy. Light can set the vanes of a toy radiometer into rotational motion. Evidently the light has mass, which strikes the radiometer. We burn fuel oil in our heating systems, and we "burn" gasoline to power our automobiles. Here again, we convert matter into energy, as we do when we burn wood to produce heat, which is a form of energy. Light is in fact the source of most of the energy available to us on Earth. It is converted by plant life into chemical energy. During photosynthesis in green plants, the energy of light is captured and used to convert water, carbon dioxide, and minerals into oxygen and energy-rich organic compounds.

Einstein had suggested that an increase in mass could be found in radioactive particles such as β particles (electrons) if they are made to travel at high speeds, and this has been verified experimentally. Another related situation is that, if one heats a mass and thereby supplies energy to it, the mass increases.

There is, fortunately or unfortunately, the converse process. A piece of matter may lose some of its mass by giving up a corresponding amount of energy. In a relatively innocuous situation one can slow up a particle, causing it to lose mass and give up energy. The unfortunate aspect is that fission and fusion of elementary particles give off radiation, and here we have the basic idea of the atomic bomb.

One key to understanding the equivalence of mass and energy is to think about how mass exhibits itself. A fundamental property of mass is inertia, the resistance to change in speed. To increase the speed one must apply energy, and the higher the speed the more energy is needed to change the speed. By the increase in speed, the body acquires more inertia or mass according to formula (1). By algebra we can show that (approximately)

$$M = m + \frac{1}{2} m \left(\frac{v^2}{c^2} \right) \qquad (2)$$

and the second term on the right side is the kinetic energy divided by c^2. Thus, the addition to the mass m is kinetic energy. Whether we say that the mass increases with the speed, that energy has mass or is mass, or that energy acts to increase mass is immaterial. The same fact obtains whether or not the increase in energy is other than kinetic. What has changed is the inertia of the energy-enriched matter.

However, Einstein went much further. It so happens that when a mass is at rest, then it is numerically true that its energy E_0 is numerically equal to mc^2, where m is the rest mass of the body. Einstein then took the relation (1) to be the mass of the body moving with velocity v; in fact, he generalized and gave an argument to show that $E = mc^2$, where E stands for the entire energy in the mass m, not just the rest mass (in our notation $E = M(c^2)$. He also showed that to a radiation energy E there must be assigned an inertia whose mass equivalent is E/c^2. These conclusions are not deductions from the special theory of relativity but are consistent with it. As Einstein put it in his book *The Meaning of Relativity,* "Mass and energy are therefore essentially alike; they are only different expressions for the same thing. The mass of a body is not a constant; it varies with changes in its energy."

In our ordinary experiences a rather artificial distinction between mass and energy is introduced. They are measured in different units, grams and ergs, and the energy E has numerically a mass E/c^2 where c is the velocity of light in the units used. However, it now seems more certain that mass and energy are two ways of measuring the same thing. If it is objected that they ought not to be confused, that they are distinct properties, we should appreciate that they are not sense-perceptible properties but mathematical terms expressing the combination of more immediately comprehensible properties, namely, ordinary mass and velocity.

Although Einstein continued to think about mechanics, electromagnetism, and other subjects, he was strongly influenced in his later work by the ideas of Hermann Minkowski (1864–1909), who was his principal professor at the Federal Institute of Technology (Zurich Polytechnic). Minkowski said in 1908:

> The view of space and time which I wish to lay before you has sprung from the soil of experimental physics, and therein lies its strength. It is radical. Henceforth space by itself, and time by itself, are doomed to fade away into mere shadows, and only a kind of union of the two will preserve an independent reality.

It is true, Minkowski agreed, that we have harbored a notion of continuously flowing time that is independent of any notion of space.

Nonetheless, when we observe events in nature we experience time and space simultaneously. Moreover, time itself has always been measured by spatial means, for example, in terms of the distance moved by the hands of a clock or by the motion of a pendulum through space. Furthermore, our methods of measuring space necessarily involve time. Even during the simplest method of measuring distance, that of applying a rod, time elapses. Hence, the natural view of events should be in terms of a combination of space and time; the world is a four-dimensional space–time continuum.

True, different observers may obtain different measures of the space and time components of the space–time interval between two events, but this is not surprising if we consider three-dimensional space itself. Two people on different parts of the globe see the same three-dimensional space, but one analyzes his experience of space into vertical and horizontal directions different from the vertical and horizontal directions of the other. Nevertheless, we continue to regard space as a three-dimensional whole rather than as an artificial combination of horizontal and vertical extents. Similarly, different observers may decompose space–time into different space and time components. This decomposition is as real and as necessary for the person who makes it as the distinction between horizontal and vertical is for a person walking down a flight of stairs. Yet we as humans differentiate; nature presents space and time together. Actually we do sometimes commingle space and time in ordinary life. We say a star is so many light-years away. The star is as far away as the distance light travels in the years of time. A train schedule is also a combination of position and time.

Einstein proceeded to utilize Minkowski's view that the universe should be regarded as a four-dimensional space–time world, but these astounding innovations of Einstein's special theory of relativity had not settled all of the difficulties enumerated in the preceding chapter. No explanation was as yet forthcoming with respect to just how gravity pulls objects to the Earth and holds a planet in its course, or why mass and weight should always have a constant ratio at a given locale.

Einstein next sought to extend the special theory of relativity to frames that are moving with respect to one another and with one *accelerated* with respect to the other. The key to a more general theory emerged in 1907 when Einstein, pondering on gravity, recognized that what is called gravitational mass is not distinguishable from inertial mass. What had caused scientists to make the distinction? According to Newton's laws of motion, when a mass is required to change direction or velocity, the mass involved in $F = ma$ is the inertial mass. Thus, if one hits a billiard ball on a table and thereby sets it in motion, the mass involved is the inertal mass. However, if one holds a billiard ball and lets it drop, it falls because the Earth's mass attracts the ball's mass. In this phenomenon, gravitational mass (weight) is involved. Are the two masses the same? This question did not trouble Newtonians, but with totally new problems arising about mass, even in the special theory of relativity, it did bother Einstein. He decided that gravitational mass was identical with inertial mass and that gravitational mass was no more than inertial mass in a totally new kind of space–time.

To pursue this thought, let us first consider a passenger in an elevator that is falling freely because the cable broke. The passenger can ignore the gravitational force, because this force does not act on the person. In fact, the passenger does not press against the floor and has no weight. If, inside the falling elevator, the passenger drops a handkerchief or a watch, they fall, but so does the elevator. Hence they remain exactly where they were dropped. Only inertial mass can act on objects *inside* the elevator. To an outside observer, however, there is a gravitational force acting on the elevator and objects in it.

Put more generally, all observations made locally on a system acted on by a uniform static gravitational force will be the same as that on a system subjected to a uniform acceleration. Acceleration and gravity are equivalent. This is Einstein's principle of equivalence. In other words, it states that an observer falling in a gravitational field will have the same experience as an observer in a region free of gravitational field but moving with an acceleration equal to that of free fall in the field.

Influenced by Minkowski's view on space–time, by his own ideas on inertial and gravitational mass, and by the desire to generalize on the special theory to allow acceleration of one frame relative to another, Einstein took up the idea of a curved space–time. The nonuniformity of a real gravitational field precludes replacing it by a single accelerated frame over a large region. Hence he used ideas of Riemann and Clifford (although he may not have known the latter), namely, that the presence of matter in space–time could be incorporated in the geometrical structure.

Einstein's four-dimensional curved space–time cannot be visualized, but an analogy may give us some intuitive feeling about it. Consider the shape of the Earth. Although for many purposes it is adequate to consider it as the surface of a sphere, it is not. There are mountainous regions, valleys, and chasms. What are the geodesics or shortest paths on this matter-filled surface? Surely they vary with the shape of the surface and vary from one region to another.

In his general theory of relativity Einstein incorporated the principle of equivalence. In this mathematical space–time, any mass will "distort" the region of space–time around it, so that all freely moving objects will follow the same curved paths, the geodesics, in that region. Classically one says the objects are accelerated because some force such as gravity acts on them. However, in the general theory the acceleration results from the properties of space–time. Hence the effect on all inertial masses will be the same, and the principle of equivalence is automatically satisfied.

Thus, the main idea of Einstein's general theory is that the geometry of space–time takes into account the presence of matter, and gravity is eliminated. (Strictly speaking all matter in space–time must be taken into account including the matter in moving objects. However, if the moving matter is small, its incorporation in the structure of space–time can be neglected. This applies to the planets.) The planets and light traveling from the sun to the Earth follow paths imposed by the structure of four-dimensional space–time. These bodies and light, freely moving—that is, not acted on by any force—follow paths that are geodesics of space–time, that is, the shortest "paths," just as in Newtonian mechanics light follows the shortest path as do other bodies not acted on by what had been attributed to the force of gravitation. Locally the space–time of general relativity is the space–time of special relativity, and generally the conclusions of special relativity carry over to general relativity.

The explanation of what were formerly considered gravitational effects in terms of the geometry of space–time disposes of another unsolved problem, namely, why the ratio of weight to mass is constant for all bodies on and near the Earth. Interpreted in the physical sense, this constant ratio is the acceleration with which all masses fall to the Earth and which, according to Newtonian mechanics, is caused by the force of the Earth's gravitational pull on the masses. Hence the constant ratio of weight to mass means that all masses follow the same space and time behavior in falling toward the Earth. However, in accordance with Einstein's reformulation of the phenomenon of gravitation, what was formerly regarded as the Earth's gravitational pull becomes the effect of the shape of space–time near the Earth. All masses falling freely must, according to the revised first law of motion, follow the geodesics of space–time. In other words, all masses should show the same space and time behavior near the Earth, and they do. Hence, the theory of relativity solves the problem of the constant ratio of weight to mass by eliminating weight as a scientific concept and by advancing an even more satisfactory explanation of the effects formerly attributed to weight.

Einstein faced one other problem. Each of us is an observer in space–time, and each of us would frame the laws of space–time in our own coordinate system. Hence, to be sure that the laws are the same for all observers, Einstein wished to frame them so that when they are transformed from the coordinate system of one observer to that of another the laws would be the same. Here Einstein faced a mathematical problem. He discussed his difficulty with his colleague, George Pick, and Pick called his attention to the tensor analysis developed by Riemann, Elwin Bruno Christoffel, Giorgio Ricci-Curbastro, and the latter's famous pupil, Tullio Levi-Civita. Einstein then found another colleague in Zurich, a differential geometer, Marcel Grossmann (1878–1936), and learned tensor analysis from him. He and Einstein wrote three joint papers in 1913–1914. In a few years Einstein was able to use Riemannian geometry and tensor analysis to formulate the theory of general relativity and to show how to transform the laws from one coordinate system to another. He acknowledged his debt to the creators of tensor analysis. Einstein wrote four papers on relativity in 1915, the crucial one of which was dated November 25, 1915. It stated that through tensor analysis the laws of nature take the same form in all mathematically acceptable coordinate systems.

The general theory of relativity was especially strange and radical for its time. What could induce mathematical physicists to accept it?

Einstein made three predictions based on his radical theories. The perihelion of a planet is the point nearest to the sun in its elliptical orbit. According to Newtonian mechanics the perihelion of the innermost planet, Mercury, should change its position from year to year by an amount that differs from the observed motion by about 5600 seconds of arc (5600″) per century. (One second of arc = 1/3600 of a degree.) A good part of this, about 5000 seconds per century, results because our observations are from a moving Earth. Leverrier in 1856 showed that part of the discrepancy, about 531 seconds per century, is the result of the gravitational attraction of the other planets. No explanation of the remaining discrepancy was successful until Einstein accounted for it in his general theory of relativity. These figures are approximate, because many new observations have been made since the prediction of 1915. Moreover, all calculations are complicated because a moving planet contributes somewhat to space–time curvature.

Einstein also predicted that light from a star would be deflected in passing the sun. As for the bending of light: light, presumably having mass, had been recognized to be attracted by a gravitational field—in this case the sun's—and scientists expected the deflection to be 0.87″ for a ray grazing the sun. Einstein predicted 1.75″. The latter figure was confirmed by observations made in 1919 during a solar eclipse. By comparing photographs of stars taken five months previously (when the stars were in the night sky far from the sun) with those taken during the eclipse, Arthur Stanley Eddington was able to show that the amount by which the starlight was deflected was consistent with Einstein's prediction (see Figure 37). Coming soon after the theory was published, Eddington's result probably did more than anything else to bring acceptance to Einstein's ideas.

Einstein predicted a third phenomenon. Atoms generally, especially those of gases when heated, emit several or even many frequencies of light. Einstein's prediction was that the frequencies emitted by atoms placed in different parts of the solar field would appear to be vibrating faster or slower than the same types of atoms on the Earth. The variations in the rates of vibration would be betrayed physically by variations in the color of the light received on the Earth. Atoms near the sun should appear redder to us on the Earth, that is, the wavelengths are shifted toward the red. This red shift has been observed.

In view of these experimental evidences, it would appear that the general theory of relativity has been amply validated. Einstein's theory contains Newton's theory as a first approximation, and this is another confirmation. However, there is one seemingly insignificant point. All the experiments we have described measure very small effects; however, Einstein was certain that he was right in his special and general theories even before they were checked by experiments.

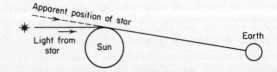

FIGURE 37

The special and general theories of relativity are now not only part of our scientific knowledge; for the phenomena involved, they offer the best knowledge we have about our physical world. Should we accept them? In particular, should we accept that simultaneity of events and that lengths and time intervals are dependent on who is observing these quantities? One could have dismissed those questions in the past, because any significant differences between two observers are dependent on very large speeds of one with respect to another. However, now that we have sent people to the moon and have sent spaceships to Saturn and on to Neptune, and that space travel will certainly expand, we no longer operate with small speeds.

In spite of the astonishing and dramatic verifications of the theory, many people find its four-dimensional, non-Euclidean universe totally unpalatable. No one can visualize a four-dimensional, non-Euclidean world, but those who insist on visualizing the concepts with which science and mathematics now deal are still in the dark ages of their intellectual development. Almost since the beginning of work with numbers, mathematicians have carried on algebraic reasoning that is independent of sense experience. Today they consciously construct and apply geometries that exist only in human brains and that were never meant to be visualized. Of course, all contact with sense perception has not been abandoned. The conclusions about the physical world predicted by geometrical and algebraic cogitations must be in accord with observation and experimentation if the logical structure is to be useful for science. However, to insist that each step in a chain, even of geometrical reasoning, be meaningful to the senses is to rob mathematics and science of two thousand years of development.

We should recall how people reacted to the fact that the Earth is spherical and later that the Earth is moving around the sun. Certainly our perception did not agree with these facts. We should then be more receptive to the relativistic concepts of time, simultaneity, space, and mass. The theory of relativity warns us against taking appearances, which hold only for a particular reference system, as the truth in any absolute sense. Here as in other physical areas mathematical laws tell us what is truth and really objective. Nature does not care much about our impressions. She continues on her course whether or not we are there.

The union of space and time and the influence of matter on space–time proposed by the theory of relativity, ideas that would have seemed outlandish to philosophers of the early 1900s, have now become embodied in a philosophy of nature more and more widely held. Nature presents herself to us as an organic whole, with space, time, and matter commingled. Humans have in the past analyzed nature, selected certain properties that they regarded as most important, forgotten they were abstracted aspects of a whole, and regarded them thereafter as entirely distinct entities. They are now surprised to learn that they must reunite these supposedly separate concepts to obtain a consistent, satisfactory synthesis of knowledge.

Aristotle first formulated the philosophical doctrine that space, time, and matter are distinct components of experience. This view was subsequently adopted by scientists and used by Newton. We, following him, have become so accustomed to thinking of space and time as fundamental and distinct components of our physical world, and separate from matter, that we no longer recognize this view of nature as man-made and as only one of a number of possible views. Of course, the philosophers of the contemporary scene, among them the late Alfred North Whitehead, do not argue that this analysis of nature is useless. On the contrary, it has proved quite valuable and even essential. However, we should be aware that it is artificial, and we should not mistake our analysis for nature itself any more than we should mistake the organs observed by dissection of the human body for the living body itself.

It is now possible to appreciate how much of science has become mathematized in the form of geometry. Since the days of Euclid the laws of physical space had been no more than theorems of his geometry. Then Hipparchus, Ptolemy, Copernicus, and Kepler summarized the motions of the heavenly bodies in geometrical terms. With his telescope Galileo extended the application of geometry to infinite space and to many millions of heavenly bodies. When Lobatchevsky, Bolyai, and Riemann showed us how to construct different geometrical worlds, Einstein seized the idea in order to fit our physical world into a four-dimensional, mathematical one. Thereby gravity, time, and matter became, along with space, merely part of the structure of geometry. Thus the belief of the classical Greeks that reality can be best understood in terms of geometrical properties and the Renaissance doctrine of Descartes that the phenomena of matter and motion can be explained in terms of the geometry of space have received sweeping affirmation.

The Dissolution of Matter:
Quantum Theory

As I have already emphasized repeatedly, no experiment has any meaning at all unless it is interpreted by the theory.

Max Born

I remember the discussions with Bohr ... and when at the end of the discussion I went alone for a walk in the neighboring park I repeated to myself again and again the question. Can nature possibly be as absurd as it seemed to us in these atomic experiments?

Werner Heisenberg

The second revolutionary development in the sciences of this century was to be called quantum theory. Up to the present time no development has altered so much our knowledge of what is real in the physical world and how nature behaves.

In discussing this subject we shall not adhere to the historical order of the discoveries, nor shall we say much about mathematical contributions or the brilliant experiments. The mathematics is quite advanced, involves several areas of the subject such as differential equations and probability, and is not readily presentable. But the reader can be assured that mathematics has as vital and instrumental a role in this area as in those we have examined earlier.

Quantum theory deals with atomic structure, and not all questions and even seeming contradictions have been resolved. We are still in the developmental stage of what is often called microphysics as opposed to macrophysics, the latter dealing generally with large-scale phenomena. Quantum theory delves far beneath what our sensations of sight and touch can possibly tell us, because only the very large atoms can be observed even with an electron microscope. The theory concerns an invisible, silent world. Though totally nonsensory in itself, its effects are as real as tables, chairs, and our own bodies. Perhaps the closest parallel is the electromagnetic world. Although we have no physical perceptions of electromagnetic waves, their effects, so to speak, are known to all of us as, for example, in radio and television.

Despite the tentative nature of some of the quantum theoretic findings, they are being applied. Atomic bombs are a reality, and they concern us now far more than some of the great mathematical creations of the past.

Whereas our sensations convince us that sound, light, water, and matter are continuous, the question of the ultimate structure of all phenomena such as light or matter had already been raised in Greek times. The belief that all matter is composed of indivisible atoms goes back to Leucippus (c. 440 B.C.) and was developed by Democritus of Abdera (c. 460–c. 370 B.C.) (The word atom comes from the Greek word meaning indivisible.) Democritus said that there are many kinds of atoms differing in size, shape, hardness, order, and position. Large bodies are composed of many atoms differing in number and arrangement, but the atoms themselves are indivisible. Both men said that all sensuous qualities are mere appearances that result from different arrangements of the atoms. Whereas shape, size, and the other variations just mentioned were physically real properties of the atoms, other properties such as taste, heat, and color were not in the atoms but in the effect of the atoms on the perceiver. This sensuous knowledge was unreliable because it varied with the perceiver.

Aristotle differed. His doctrines, which originated with Empedocles (490–430 B.C.), maintained that there are four elements, the essences of which appear in earth, fire, air, and water, and that all objects possessed more or less of these essences. These essences combined under attraction (love) and repulsion (hate) and explained all material appearances. Actually, other elements (for example, copper, tin, and mercury), were known in Greek and pre-Greek times, but these were not analyzed by Aristotle and his successors. Aristotle also believed that even atoms were divisible, in fact infinitely divisible, so that matter was continuous and there were no ultimate individual particles. Aristotle's views dominated until the sixteenth century.

However, from the seventeenth century until the beginnings of this century the generally accepted theory was that atoms were indivisible. There were supposedly different atoms for basically different elements such as hydrogen, oxygen, copper, gold, and mercury. Moreover, it was accepted that, while the atoms of any one element have the same weight, atoms of different elements differ in weight. Ordinary material substances such as water are composed of molecules that are combinations of atoms. The beginnings of chemistry were undertaken on this basis, notably by Robert Boyle (1627–1691) in his *Sceptical Cheymist* (1661).

A more positive declaration, in agreement with Boyle's, was made by John Dalton (1766–1844) in 1808. Dalton's major idea was that many of the laws of chemistry could be explained easily if it were assumed that to each chemical element there corresponded specific atoms of matter. Every substance is made up of different combinations of different kinds of indivisible atoms.

By 1860 about sixty different types of atoms were known. In that decade Dmitri Ivanovich Mendeléev (1834–1907) undertook to arrange the known elements in the order of their atomic weights. He noted that each eighth element among the first sixteen had similar chemical properties. He found, however, that beyond this point, if he were to continue to arrange elements in the order of increasing atomic weight and yet put elements with similar chemical properties eight positions apart, he had to leave some blank spaces. It seemed a fair inference to Mendeléev that unknown elements belonged in the blank spaces. This reasoning led Mendeléev to look for new elements, and soon investigators found three, now called scandium, gallium, and germanium, whose chemical properties Mendeléev could and did predict from a knowledge of the properties of the elements eight positions below. Later discoveries caused some modifications in Mendeléev's periodic law, but his arrangement still is the essence of the modern one. Although Mendeléev recognized that he had no physical explanation of why the regularity revealed by his arrangement held, he advocated utilizing the periodicity to discover new elements, to determine their atomic weights, and to predict the chemical properties of these elements such as their ability to combine with other elements to form molecules.

The elements found by Mendeléev and later investigators have been ordered, and they are numbered on the basis of simplicity of structure. Thus hydrogen is number 1. Helium is number 2, and so on up to 103 for Lawrencium. The atomic weights of these elements are the number of times heavier an element is than the hydrogen atom. The weights are 1 for hydrogen, 4 for helium, and up to 257 for Lawrencium.

Although the issue of indivisibility of atoms versus continuity was still occasionally debated, up to 1900 atoms were generally accepted as indivisible and the ultimate constituents of matter. In 1907 Lord Kelvin said the atom was indestructible. However, several remarkable discoveries refuted the doctrine of indivisibility. In the 1870s the thought gained ground that the atom might be a composite of particles. In 1897 Sir Joseph John Thomson (1856–1940) gave some evidence that the atom does consist of particles and obtained fairly good values for the electric charge and mass of very light electrified particles called electrons. In 1900 Hendrik Antoon Lorentz (1853–1928) agreed that these negatively charged particles exist. The mass of such electric particles was found to be about 10^{-27} grams, roughly 2000 times less than the lightest atom, the hydrogen atom. The charge of an electron is minute, about 4.80325×10^{-10} electrostatic units. Thomson proposed in 1904 the atomic model of a nucleus surrounded by these smaller electrons. This was the first break with the traditional belief of an indivisible atom.

At this time atomic theory was fairly simple. All atoms consisted of protons (which are positively charged) and electrons. The protons were taken to constitute the nucleus of the atom. It soon became clear that the mass of the atom is almost entirely concentrated in the nucleus. The hydrogen nucleus is the smallest, and its mass is 1.6724×10^{-24} grams. Surrounding the nucleus of any atom are the electrons, the number of which is the atomic number.

Another break with traditional theory came when rather accidentally Antoine Henri Becquerel (1852–1908) discovered radioactivity in 1896. The subsequent study of this phenomenon was carried out by two members of the Curie family, Pierre (1859–1906) and Marie (1867–1934). It became somewhat evident that the atom is a far more complicated structure than had thus far been envisioned. Although we shall say more later about the nature of radioactivity, what was soon clear to the discoverers was that the nuclei of some atoms, such as the very heavy ones, emit particles and electromagnetic rays, which became known as alpha, beta, and gamma rays. The alpha particles are ionized helium atoms; the beta particles are electrons; and the gamma rays are very-high-frequency electromagnetic waves. Moreover, when an alpha particle is emitted from an atom, the atom is converted into one of a somewhat lighter element. In the early work on atomic structure the products of radioactivity were used to study atomic particles.

By 1910 Ernest Rutherford (1871–1937), who had experimented with radioactive atoms, conceived the idea that atomic structure resembled that of the solar system in which there is the centrally located sun surrounded by the moving planets. In Rutherford's pro-

posed atom there was the centrally located nucleus surrounded by electrons that moved in different orbits. He was quite sure that the volume
of the nucleus was about one-billion-billionth of the atom as a whole
(10^{-12}). Thus gold, for example, which has the atomic number 79, has
a nucleus that is surrounded by 79 electrons. The nucleus, as noted,
consists primarily of particles called protons. However, to account for
the atomic weight Rutherford suggested that the nucleus also contains
electrically neutral particles, which he called neutrons. There are nuclei
with the same number of protons but a different number of neutrons;
such atoms are called isotopes.

While men such as Rutherford were discovering and formulating
what they believed to be the structure of the atoms, a vital discovery
that affected all later work in atomic theory was made by Max Planck
(1858–1947) in 1900. Planck was concerned with what is called ther-
mal radiation, or black body radiation. For example, red-hot metal
emits light, which, as we know, is one type of electromagnetic radiation. For good reasons, though not mathematically based, Planck
declared in 1900 that the radiation is emitted not in one continuous
"stream" but in little packets or quanta that are smaller or larger
depending on the frequency of the radiation normally emitted by the
atoms in question. Planck's formula for the energy in radiation is

$$E = nhf$$

where n, the number of quanta emitted, can be 0, 1, 2, ... , h is a
constant now called Planck's constant (about 10^{-26}), and f is the frequency of the radiation, which is a composite of all the quanta somewhat as water waves are a composite of the molecules of water. Radiation such as light does seem continuous, but this is because the number
of quanta radiated by an ordinary electric light bulb is so great: 10^{20}
quanta per second for a 100-watt bulb.

Conversely, light of frequency f falling on a metal surface liberates
energy. The law of emission tells us that the energy of each emitted
electron is proportional to hf. These quanta were later called photons.
Planck's formula was an assumption, a lucky guess, or a remarkable
intuition. However, Planck used a good deal of mathematics to represent and deduce many of his conclusions.

Einstein's paper of 1905 on the photoelectric effect, the details of
which we need not examine, not only confirmed Planck's formula but
used it to good advantage. As Planck had concluded, light shining on
metals causes electrons to come out of metals. The incident radiant

energy is quantized. It consists of quanta, each of amount hf. Moreover, the energy of each emitted electron is proportional to hf. Only by postulating quanta could Einstein explain the interaction of light and atoms. It takes place only for very high frequencies but does not depend on the intensity of the light. However, the number of electrons expelled does depend on the intensity. Planck's and Einstein's work raised the problem of whether light and all radiation consists of waves or particles. We shall say more about this later. What may be clear at this point is that electromagnetic radiation exhibits both wave-like and particle-like behavior.

Let us return now to the work on the structure of atoms. Rutherford's theory of the atom failed to account for the fact that electrons moving around the nucleus did not radiate light or some energy, in accordance with electromagnetic theory, and ultimately spiral toward the nucleus. Niels Henrik David Bohr (1885–1962) "looked" a little more closely into the structure of the atom, and although he accepted Rutherford's solar system model he decided, using mathematical theory, that the electrons in atoms should not radiate energy just because they are in motion, but that electrons can revolve only in certain definite orbits as the planets do. A revolving electron possesses energy, namely, the mechanical energy that any revolving object possesses. However, only when an electron moves from one orbit to another can it emit or absorb radiation. Moreover, the emission and absorption are effected in jumps that are quanta—that is, multiples of the quantum hf. When an atom absorbs radiation, an electron moves from an inner to an outer orbit, and in the converse motion, it emits quanta or photons.

The Bohr theory did not explain all the observations about frequencies of radiation emitted by the atoms, and so work on the structure and behavior of atoms continued.

Thus far it was radiation that was conceived of as quanta or photons that acted like particles. In 1922 Louis-Victor de Broglie (1892–   ) entered the picture and posed an idea that became the central theme in what is now known as wave mechanics. De Broglie considered the particle nature of light waves (the photons) and asked: If light waves behave like particles and waves, might not all particles have wave properties? More generally, should not one consider waves associated with all matter? The waves should have a frequency and a velocity.

With the aid of some mathematics in the area of partial differential equations, de Broglie inferred that the wavelength (λ) of the wave associated with a particle should be equal to Planck's constant h divided by the product of the particle's mass m and its velocity v. Specifically,

$$\lambda = h/mv$$

The product mv is the momentum and is usually denoted by p. For a mass of one gram and a velocity of one centimeter per second, $\lambda = h$, which is about 10^{-26} centimeters—ten million times smaller than an atomic nucleus. Hence, in the familiar large-scale world of matter, all objects are of enormous size compared to their matter waves, and so we do not observe matter waves.

Erwin Schrödinger (1887–1961) followed up on de Broglie's idea that waves are associated with all material particles, particularly electrons, and formulated in 1926 a partial differential equation for a function ψ that would express the shapes of these waves. Solving the equation gives the shapes; the solutions are called eigenfunctions or characteristic functions. When numerical values are given to the constants in the solution, the functions lead to particular values of the solution called eigenvalues or characteristic values. The discrete energy values of the electrons in an atom appear as the eigenvalues of the wave equation, and they agree with Bohr's theory insofar as they are obtainable.

To get some idea of Schrödinger's concept of how an electron wave behaves, let us consider a rough picture. Figure 38 illustrates a wave that in the present case extends over two wavelengths. If this wave were suitably bowed on a violin string, it would vibrate up and down and take all positions between the continuous and broken-line curves. It might also vibrate in a sequence of wavelengths that are fractional parts of the fundamental wavelength (for example, one-half, one-third). In Schrödinger's case the total electron wave of any one electron surrounds the nucleus and may extend over two, three, or

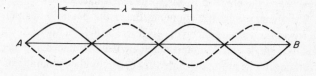

FIGURE 38

even five fundamental wavelengths. In each case there is an integral number of wavelengths, and the end of the final wave joins the initial wave. In terms of Figure 38, point B would meet point A.

Schrödinger's ψ represents the magnitude of his matter waves, which vary from point to point in space and from moment to moment in time. They are standing waves restricted essentially to a small region about the nucleus; each trails off gradually with increasing distance from the nucleus, but most of the waves lie within the region known experimentally to be the sizes of the respective atoms. Thus, for the hydrogen atom in its lowest energy state, the wave pattern has appreciable magnitude only within a sphere of about 10^{-8} centimeters. For any one type of atom the solution of the Schrödinger wave equation gives the discrete wave patterns of the electrons, and with each of the states it associates a particular value of the energy.

To repeat, the Schrödinger wave describing an electron in an atom must be thought of as containing a range of different wavelengths, rather than the single wavelength of a simple wave form. It is similar in this respect to the composite sound waves produced by a musical instrument.

An obvious question about the de Broglie-Schrödinger waves is: What they are made of, or put differently, what substance does the wave consist of? Such a question was also posed in the nineteenth century when light waves and other electromagnetic waves were discovered. Physicists initially considered those waves to be vibrations of a mysterious substance called ether and made various mechanical models of its action. Eventually such ideas proved untenable, however, and the waves were regarded as independent entities. In the case of electron waves, Schrödinger at first suggested that the waves actually represented the distribution of the electron's charge, so that in an atom the charge of the electron and the density of the electron were physically distributed over the region of space in which the wave is different from zero. However, this behavior is never observed. Instead, whenever the electron is found, its total charge is localized in a small region of space, and the electron is of a corpuscular nature.

Strictly speaking, the enumeration of the possible distinct wave patterns in the various energy levels refers to those associated with the states of a single electron considered by itself. When many electrons are present in one atom, their identity becomes blurred and their wave patterns merge into a single one associated with all of them in common.

The Schrödinger pictures of the electrons are like clouds of varying density; they are three-dimensional. The clouds surround each
other and each varies in intensity from zero to a maximum and then
to zero. The clouds extend beyond the atom but are most dense for
each electron at the distance from the nucleus predicted by Bohr. The
cloud, an interpretation of the mathematics, is a picture that is una-
voidably unclear. It is not possible to visualize exactly Schrödinger's
mathematics. Finding analytical solutions of Schrödinger's equation is
so difficult that only a few problems have been solved exactly. Yet
these few are in excellent agreement with experiments, and others,
though approximate, are also seemingly in agreement with experiments. One of the problems completely solved is that of atomic hydrogen; it answers every question for which there is an experimental
check.

That electrons do behave as waves in some circumstances was
demonstrated experimentally in a famous experiment performed by
Clinton J. Davisson (1881–1958) and Lester Germer (1896–1971), and
by George P. Thomson (1892–1975) in 1927. They demonstrated diffraction of electron waves (using the lattice structure of crystals). Diffraction is the phenomenon of waves bending around an object that is
struck by the waves. It is in principle what happens when water waves
bend around the stern of a ship. Hence certainly in some phenomena
particles behave as waves. Physicists are now convinced that all subatomic particles have waves associated with them and that wavelengths satisfy the relationship deduced by de Broglie. Thus, the work
of de Broglie and Schrödinger brought to the fore the troublesome concept of particle–wave duality.

Despite the evidence that electrons do in some circumstances
behave as waves, the idea that electrons are "smeared out" around the
nucleus of an atom was not acceptable to all physicists. Because the
electrical charge of an electron is a definite fixed quantity, the notion
that in any small region the charge density must be infinitesimal was
objectionable. Charges were always multiples of the electron charge.
For this reason, and to avoid the wave–particle duality, a totally
different interpretation of the Schrödinger theory was proposed by
Max Born (1882–1970) in 1926. Born introduced the probabilistic
interpretation.

The theory of probability, which entered into mathematics quite
by chance in connection with games of chance, had already been used
in the latter part of the nineteenth century by Maxwell and Ludwig
Boltzmann (1844–1906) to study and arrive at laws that would

describe the motion of gases—the kinetic theory of gases. Indeed, one of Einstein's famous papers of 1905 was devoted to this topic, which has been called Brownian motion. Instead of regarding an electron as spread out in a cloud, the density of which varies from point to point, Born interpreted the density to be the probability of finding the electron as a particle at any point.

Referring to the function ψ involved in Schrödinger's differential equation, Born suggested that $|\psi|^2$ gives the probability that a particle is in an element of volume of space and instant of time. Thus, electrons as particles are known to be somewhere only to a degree of probability. For example, if $|\psi|^2 = 0.8$ at some point, then the probability of finding the particle (electron) in a small volume enclosing that point is about 80 out of 100. The probability interpretation is still the accepted one.

The probabilistic interpretation yields an accurate estimate of the probability that an electron will be found in any given volume. When detected, it is not smeared out, contrary to the matter–wave theory of Schrödinger. However, there is the question of whether the probabilistic interpretation is the best possible one or just partial ignorance.

The use of probability may seem to be a last resort, but the history of what is called statistical mechanics shows its value. Any gas is a collection of chaotic molecular motions. However, the pressure of a gas and other quantities can be calculated by using the most probable values, and these are physically highly significant.

Einstein, Planck, and Schrödinger objected to the probabilistic interpretation. Einstein stated his objections in a paper of 1935. His argument was that quantum theory is approximate and incomplete:

> I reject the basic idea of contemporary statistical quantum theory, insofar as I do not believe that this fundamental concept will prove a useful basis for all of physics. . . . I am in fact firmly convinced that the essentially statistical character of contemporary quantum theory is solely to be ascribed to the fact that this theory operates with an incomplete description of physical systems.

Although the probabilistic interpretation is now accepted, perhaps with further investigation one might be able to determine the precise location of the electron with certitude. However, according to one of the novel features of quantum theory, some indeterminism is unavoidable. This is the principle of uncertainty enunciated by Werner Heisenberg (1901–1976) in 1927. Roughly stated, it asserts that we can never expect to have accurate information about both the position and velocity (or momentum) of a particle at any given instant of time. More precisely, Heisenberg showed that the product of the uncertainty in position and momentum must be at least $h/2\pi$. He was certain of this uncertainty and ascribed it to the fact that particles are both waves and corpuscles. One can measure position or momentum separately and precisely but not both at the same time. Heisenberg also said that in such fine measurements the probing object becomes relevant.

The reason for the latter cause of uncertainty is that the measuring apparatus is not smaller or finer than the electron. One can use only other electrons or photons, but these in themselves have an intense effect on the particles under observation. Hence, we cannot observe an event in the atomic world without disturbing it. Because we cannot know position and velocity precisely at any moment, we cannot predict much about the particles. We can predict probabilities. The observations and experiments of classical physics will not serve.

If Planck's constant were large, this uncertainty would apply to macroscopic phenomena and one could not be certain, for example, that a gun aimed by a perfect marksman at a target would strike it. But quantum mechanical reality does not correspond to macroscopic reality. Uncertainty is built into wave mechanics. The uncertainties in position and momentum, however, are very small, and their effect on observable (macroscopic) phenomena is negligible.

The uncertainty principle of quantum theory also refutes the classical idea of objectivity—the idea that the world has a definite state of existence independent of our observing it. This contrasts with our ordinary experience of the world, which supports the classical view of objectivity that the world goes on even if we do not perceive it. Wake up in the morning and the world exists much as you left it. The uncertainty interpretation, however, maintains that if we look closely at the world—at the level of atoms—then its actual state of existence depends in part on how we observe it and what we choose to see. Objective reality must be replaced by the observer-created reality.

Further investigation of the structure of atoms began to concentrate on the nucleus. Of course, radioactivity, as we have noted, gave some indication that the nucleus is not an indivisible particle. Radioactive atoms give off alpha rays, which have a positive charge double that of the electron and a mass four times that of the hydrogen atom; beta rays, which are essentially electrons; and gamma rays, which have the highest known frequency of electromagnetic waves. All emanate from the nucleus of heavy atoms.

Further experimental investigation of the nucleus of atoms, mainly by the use of "accelerators" or atom smashers, soon revealed that the nucleus is definitely not a single entity but contains a variety of distinct particles, including protons, neutrons, (which are essentially uncharged), positrons (which were discovered in 1933 and have a positive charge), leptons, mesons, baryons, hadrons, pions, neutrinos, quarks, and many others. New ones are constantly being discovered; that is, their existence is inferred from experimental effects. There are relationships among the many nuclear particles, but for our purposes it is sufficient to note their presence.

Despite the variety of nuclear particles, protons and neutrons are the major building blocks of all matter. Ninety-nine and nine-tenths percent of our bodies are attributed to them. All nuclei heavier than hydrogen contain neutrons in addition to protons.

Some of the elements of the nuclei also exhibit wave properties much as electrons do. This is particularly true of the hydrogen and helium nuclei. It is equally true that, as regards their mechanical effects, all nuclei retain the character of particles.

What is more surprising about this variety of particles inside and outside the nucleus are the changes that take place in the structures of the particles. For example, a proton can give up its one unit of positive charge to a neutrino; the proton thus becomes a neutron and the neutrino becomes a positron, which has the same mass as an electron but a positive charge of the same numerical value as that of the electron. The existence of positrons was predicted by Paul A. M. Dirac's (1902–1984) theory of 1932. Conversely, a neutron can emit an electron and a neutrino; the neutron then becomes a proton.

A quantum or photon can be "split" into an electron and positron. A positron and an electron may combine to form two or more photons. We have here one of the examples of the conversion of mass to energy and vice versa.

Thus, not only are nuclei not indivisible, but they keep changing. Many particles also decay, some quickly and some very slowly, and become converted to radiation. However, protons and electrons are not known to decay, although protons may decay over 10^{30} years. They are presumably composed of quarks, which in turn change to leptons. The conversion of a quark to a lepton may produce proton decay. Experiments to detect proton decay are now under way.

To make matters more complicated, quantum theory has led us to a very unique phenomenon—antiparticles—discovered in the 1930s and predicted by Dirac. For many particles scientists have isolated antiparticles of opposite electric charge and the same mass. It is believed that if a particle meets an antiparticle, they annihilate each other and produce a particle of lesser mass. We have already noted that the combination of electron and positron produces two or more photons. Proton and antiproton, which are of the same mass, annihilate each other and produce mesons, some of which become photons. There are also antineutrons, antineutrinos, antimesons, and antiquarks. The total number of particles and antiparticles is now about eighty. Whether there is an antiparticle for each type of particle is not known. One would hope that the variety and number of antiparticles is limited; else we shall all become radiant energy! Fortunately, there is surprisingly little antimatter in the universe.

We have said nothing about the forces that act between particles. What keeps protons together? Their positive charge should cause them to repel each other. Beyond the already familiar gravitational and electromagnetic forces, physicists have postulated that there are both strong and weak forces that keep the protons and neutrons together. Not much is known about the nature of these two forces, but investigation of their nature is under way.

What then are we left with in the way of a cohesive vision of this microcosmic universe? The concepts and conclusions of quantum theory seem outrageous. They defy, deny, or at least, offend common sense. Before we seek to mollify our initial reactions, let us first note how seriously we must consider the reality of the theory. We know that the full theory of atomic structure explains much of molecular structure and a good deal of chemistry. However, there are more real or at least more visible realities: atomic bombs created by the processes of atomic fission and fusion. If an atom of uranium is hit by a neutron, it splits and part of its mass is converted to an enormous amount of energy. Moreover, the process can be made to perform in a chain reaction. This is the basic idea of the atomic bomb and nuclear power. And we have seen this reality.

The opposite process takes place in fusion, which is still to be harnessed. If four nuclei of ordinary hydrogen atoms are fused into one helium atom whose atomic mass is slightly less than four (3.97) times the hydrogen mass, immense amounts of energy are released in the form of light and heat. This process is constantly taking place in the sun. On the Earth we must use isotopes of hydrogen—namely, deuterium and tritium—whose masses are two and three times the mass of the basic or light hydrogen, and fusion of these requires enormously high temperatures.

It is interesting that Sir Oliver Lodge predicted in 1920 that "the time will come when atomic energy will take the place of coal. . . . I hope that the human race will not discover how to use this energy until it has brains enough to use it properly." Yet Rutherford said in 1933 that atomic power was absurd.

Our survey of quantum mechanical processes has shown that matter in the sense of particles can be converted into waves, and vice versa. What, then, is the physical reality? To answer this question Bohr proposed the theory of complementarity, a duality in which there are, he said, neither pure waves nor pure particles in nature, but only entities that partake of both characters. The photon is not a wave in the old sense. It is a particle-wave, and it combines features of both aspects. Nor is the electron a particle in the old sense; it is a wave-particle. Whether the entity photon or the entity electron behaves as a wave or as a particle depends entirely on the experiment we perform on it. Neither entity ever behaves at the same time as both a wave and a particle. Thus, if we do an optical-interference experiment with light, then our photon behaves as a wave. If we do a photoelectric experiment with the same light, the photon behaves as a particle. When we use electrons in radio tubes or in television tubes they behave as particles, but curiously enough, if we shoot a beam of electrons through a crystal we then get interference effects, exactly as with light waves. As Heisenberg put it in his *Physics and Philosophy,* "What we observe is not nature itself, but nature exposed to our method of questioning."

Is there then any significant difference between the particle-wave and the wave-particle? The answer is, very much so. The most fundamental distinction between them is again optical. Whereas the wave-particle can never have the velocity of light (otherwise its mass would become infinite), the particle-wave, being a light wave, has precisely this velocity. There are many other differences. The particle-wave photon cannot have what we call rest-mass (i.e., weight when not moving), for if it had it would have infinite mass when traveling as light. The wave-particle has, of course, its natural rest-mass. There are still other differences.

Wave and particle are concepts borrowed from everyday life. Are they appropriate? Or does the decision depend on what apparatus we use to observe a phenomenon? Perhaps we need new concepts.

As to particles versus waves, some contend that matter is not as solid and enduring as it has seemed to be. Instead, it must be thought of as energy in a highly concentrated state but prone to explode into massless particles flying off into space with the speed of light. Furthermore, the continuous fields we had accepted, such as the electromagnetic field, are largely mathematical conceptions though with strong overtones of reality. In fact, some philosophers of science and physicists assert that fields alone are real, that they are the substance of the universe, and that particles are merely the momentary manifestations of fields. Light waves are the overall mathematical pattern of photon behavior. They are, to use Einstein's words, ghost waves. Schrödinger in a paper of 1926 asserted that waves are the only reality, and that particles are only derivative concepts.

A more extreme view is that substance, in its traditional meaning of permanent, divisible, corpulent, hard, and extended, has dissolved in our hands and no longer exists. All we have are some total of mass and energy. The total is conserved, but one may be converted into the other. Thus in some particle actions, where a particle is used to hit another in an accelerator, new particles are formed, and among these are the particles that were there to start with. How is this possible? The energy imparted to the projectile particle becomes mass. Energy is mass according to $E = mc^2$. Whenever energy is present, mass is also present. They are complementary aspects of a reality that does not lend itself to pictorial representation. Any macroscopic description does not get to the behavior of microscopic phenomena, and any attempt to answer the questions that are familiar in Newtonian physics will not work in atomic phenomena.

If all matter reduces to quanta or particles, why do we not notice these in our everyday experience? The answer is that even a particle of dust is enormously more massive than an atomic particle. Even if the particle were given an unreasonably small velocity, the wavelength would be so small that no effects of quantization of motion would be noticeable. We are justified in saying that as we leave the atomic realm and approach the realm of ordinary events, the quantum concepts merge into the classical ones. These are apparent in the medium-sized world but not in the atomic or even cosmological worlds.

Quantum theory predicts very well the results of experimentation. However, understanding of the physical processes is still not adequate. Thus, an electron is represented mathematically by a wave function that describes the electron as smeared out in space. This wave function does give the probability that the electron will be found at any given place—when found, it is not smeared out but has a definite position. Hence, what is the correct picture of the electron? The pointer positions of the instruments are correctly predicted, but the underlying physical happenings are not clear. The mathematical rules work, but a commonsense interpretation of this world is sadly lacking. Apparently both waves and particles are needed to describe reality.

The order of the universe may be the order of our own minds. We are not merely the observers of reality; we are also the participants. Nature is not an open book that we as independent observers can read. This renunciation of accustomed demands on physical explanation has caused many physicists and philosophers to doubt that we have a suitable description of atomic phenomena. In particular, the probabilistic mode of description should be regarded as a temporary expedient to be replaced by a deterministic description.

We must, however, keep in mind that quantum theory is a relatively new development. Perhaps in another fifty years the confusion of particles and particle–wave theory will all be resolved in a remarkably simple theory. Much of our knowledge of the variety of particles is inferred from strings of "dots" on a cathode-ray screen. These dots appear when particles are bombarded in modern accelerators. On the other hand, these accelerators impart enormous amounts of energy to the bombarding particles, and one might infer that this energy is converted into mass. Is matter truly real or just an impression derived from our unreliable and superficial sense observations? On a large scale, mass is surely a statistical effect.

We see that an understanding of atomic structure has major value for physics, but it is also invaluable in aiding chemical and biological research. Perhaps biochemistry will disclose the secrets of life and heredity, and in so doing improve our health and extend our lives. At least we can say that our explorations into the nature of the atom have been fruitful.

What is most relevant for us to see is that our models of atomic structure are not physical. They are entirely mathematical. The mathematics provides order where there was chaos. As Dirac and Heisenberg put it, a consistent mathematical description of nature is the road to truth in physics. The need to visualize or to obtain a physical account is a holdover from classical physics.

The Reality of Mathematical Physics

We are in a position similar to that of a man who was provided with a bunch of keys and who, having to open several doors in succession, always hit on the right key on first or second trial. He became skeptical concerning the uniqueness of the coordination of keys and doors.   Eugene Wigner

We began our subject by asking whether there was an external world, and we answered, despite contrary assertions by Berkeley and qualified assertions by other philosophers, that there is one. However, we did show that our sense perceptions not only are limited but also can be misleading. Nor does intuition even when sharpened by experience prove to be much more helpful. Thus, despite the rather artificial character of mathematics, we did resort to it to correct and extend our knowledge of the external world.

Hence, humanity accepted that the Earth moves around the sun, not because the heliocentric theory is more accurate than the older geo-centric theory, but because it is mathematically simpler. Certainly the heliocentric theory is from a sensory point of view less credible.

To account for the motions of the planets in their orderly elliptical paths, Sir Isaac Newton provided a theory of gravitation whose phys-ical nature neither he nor his successors for three hundred years have explained. Sense perception in this case has proved useless.

Purely mathematical considerations led Clerk Maxwell to assert that there are electromagnetic waves that are not at all detectable by any one of the five human senses. Yet the reality of these waves can hardly be questioned; every radio and television set proclaims their reality. Maxwell also concluded that light is one type of electromag-netic wave. In this case, one could say that a mystery was replaced by mathematics.

197

The inability to find an absolute frame of reference for space and time, despite Newton's belief in such absolutes, and the desire to reconcile the laws of Newtonian mechanics and the laws of electromagnetism led Einstein to create the special theory of relativity. His theory, somewhat loosely put, states that length, mass, time, and simultaneity are entirely relative to the observer. Experimental confirmations compel us to accept these conclusions as factual. That the general theory of relativity dispenses with the mysterious force of gravity lends more than credence to this theory. Moreover, the fact that experimental findings were predicted by the general theory adds to our conviction that the theory is sound.

The theory of atomic structure, quantum theory, almost challenges us to disbelieve it. All of it deals with phenomena we cannot observe directly but whose nature we must infer from effects. That electrons when ejected from atoms behave not as particles but as waves, and that one can also interpret the behavior of the ejected electrons as particles that have only a probability of being in any one location at a specific time certainly strains our credulity. That the nucleus of an atom contains many dozens of particles and antiparticles, many of which decay gradually, leaves us with the incredible conclusion that essentially there is no solid matter in the external world and raises the question, of what substance are we human beings composed. Admittedly quantum theory is a relatively new branch of science, and perhaps we should regard all of its conclusions as tentative. Still, atomic bombs and nuclear energy are realities.

The chapters in which we have surveyed mathematical physics do not, of course, cover all its achievements. A subject such as hydrodynamics treats the mathematical behavior of water, gases, and other liquids, but it does not proffer surprises about reality. We have at least some physical awareness of that with which it deals. However, this is not true of the phenomena we have surveyed. These either deny our sensory knowledge or are totally lacking in sensory knowledge.

What we accept as reality today is in striking contrast to what earlier generations accepted—whether Aristotelians or the seventeenth- and eighteenth-century mathematical physicists. As the laws of motion and gravitation extended their sway over more and more phenomena, and as planets, comets, and stars continued to pursue paths so precisely described by mathematics, the assumption of Descartes, Galileo, and Newton that the universe is interpretable in terms of matter, force, and motion became a conviction in the minds of almost every thinking European.

Berkeley once described the calculus concept of the derivative as the ghost of departed quantities. Much of modern physical theory is the ghost of matter. However, by formulating mathematically the laws of these ghost fields, which have no apparent counterparts in reality, and by deducing consequences of these laws, we obtain conclusions that, when suitably interpreted in physical terms, can be checked against sense perceptions.

The fictional character of modern science was emphasized by Einstein in 1931:

> According to Newton's system, physical reality is characterized by the concepts of space, time, material point, and force (reciprocal action of material points. . . .

> After Maxwell they conceived physical reality as represented by continuous fields, not mechanically explicable, which are subject to partial differential equations. This change in the conception of reality is the most profound and fruitful one that has come to physics since Newton. . . .

> The view I have just outlined of the purely fictitious character of the fundamentals of scientific theory was by no means the prevailing one in the 18th and 19th century. But it is steadily gaining ground from the fact that the distance in thought between the fundamental concepts and laws on one side and, on the other, the conclusions which have to be brought into relation with our experience grows larger and larger, the simpler the logical structure becomes—that is to say, the smaller the number of logically independent conceptual elements which are found necessary to support the structure.

Modern science has been praised for eliminating humors, devils, angels, demons, mystical forces, and animism by providing rational explanations of natural phenomena. We must now add that modern science is gradually removing the intuitive and physical content, both of which appeal to the senses; it is eliminating matter; it is utilizing purely synthetic and ideal concepts such as fields and electrons about which all we know are mathematical laws. Science retains only a small though vital contact with sense perceptions after long chains of mathematical deduction. Science is rationalized fiction, rationalized by mathematics.

Science now defines a dynamic kind of reality, one that grows and changes as our understanding grows and changes, however much it tries to incorporate what is truly permanent. Moreover, we must admit reality to objects and phenomena that are not directly perceptible. Sensory confirmation is not required. Nature is richer than our senses tell us. There are no commonsense counterparts for atoms, electrons, curved space–time, or electric fields. The lack of a meaningful model is a great disadvantage, of course, particularly to people who are bound by commonsense experiences and who quite naturally tend to reason on the basis of these experiences.

In humanity's efforts to determine what is real, a new factor, the role of the observer, has entered. In the nineteenth century nature still appeared as a set of phenomena in which man and man's intervention in nature could be ignored in principle, if not in practice. In our century, however, it is especially in the sphere of atomic physics that we have had to abandon the old view. The reality of elementary particles is not clear. What we do know are processes, bubble chambers, photographs, and "television" displays of effects, by which we infer what seems to be the behavior of elementary particles. These processes are as much involved in the understanding of the behavior as the underlying behavior. Moreover, as Heisenberg pointed out, the processes we use affect the true physical behavior. Where classical mechanics appealed to either a particle picture or a wave picture, there was an objective reality independent of the observer. Today the laws of physics concern our knowledge rather than what may be true in the physical world. We observe, manipulate, mathematize, and draw our conclusions.

Classical physics allows for errors in measurement and so uses statistical theory and probability. However, the quantity measured is fixed, and there is an exact value. This is not the case in quantum mechanics, where the happenings are known only statistically. There are no measuring devices of more and more refinement as there are in classical physics. The existence of the particles in space and time is inferred. Thus, the objective reality of the elementary particles has been strangely dispersed, not into the fog of some new ill-defined or still unexplained conception of reality but into the clarity of a mathematics that no longer describes the behavior of the elementary particles but our knowledge of this behavior.

We have therefore come to accept that the real world is not what our unchallenged senses tell us or what our limited perceptions enable us to say but rather what man's major mathematical theories tell us. In the case of Euclidean geometry, although the concepts of point, line, plane, and the like are idealizations, they are idealizations of real objects and one can point to real points, lines, and planes as the reality. What should we point to in the cases of gravitational force and electromagnetic waves? We observe their effects. But what is physically real beyond the mathematics? Not even physical pictures, admittedly imaginative, suffice to explain the nature of these forces and fields. It seems impossible to escape the conclusion that *mathematical knowledge is our only grasp of some parts of reality.*

Is even mathematics real in the sense that what it says about the real world is actually physically real? Perhaps by considering some application of mathematics we can answer this question. Johannes Kepler triumphantly announced that each planet moves in an elliptical path around the sun. But was the ellipse *just* exactly what he really needed? The answer is no. Kepler spent years in trying to fit a curve to the astronomical data on the path of Mars, and he thought of the ellipse because it was already known in mathematics. When he found that the elliptical path did fit the observations well enough that the departures from that path could be charged to experimental error, he decided that the ellipse was correct. However, the paths of the planets around the sun are not ellipses. If there were just the sun and only one planet in the sky, and if each could be regarded as a perfect sphere, then the path of that planet would be truly elliptical. But the actual gravitational pull on any one planet is not that of the sun alone. It includes the pull of all the other planets and many moons. For this reason, the path is not elliptical. It was lucky that Tycho Brahe's astronomical observations, which Kepler used, though superior to all preceding ones, were still crude enough to allow Kepler to consider an ellipse a good fit.

What about Einstein's use of Riemannian geometry and tensor analysis? Were Riemannian geometry and tensor analysis just right for the theory of relativity? Most likely not. There is ample reason to believe that Einstein just did the best he could with the mathematics he found available. Ingenious as the general theory of relativity is, it is contrived; it is not very useful for the solution of astronomical problems because it is too complicated; and the evidence for it rests solely on the improved accuracy with which it predicts only three astronomical phenomena. If the history of science teaches us anything, it teaches us that this theory will some day be superseded.

These two examples tell us clearly that what mathematics says is not necessarily just right about the physical world. Nature neither prescribes nor proscribes any mathematical theory. Mathematical physics must also employ physical axioms such as Newton's law of gravitation. The axioms may indeed seem to be generalizations of experience, but such generalizations may be somewhat erroneous. Predictions confirmed by experimental verification to justify the mathematical and physical axioms must be used cautiously. This was emphasized by Bertrand Russell in his *The Scientific Outlook* (1931). He gave the following example. If one starts from the assumptions that bread is made of stone and that stone is nourishing, one can logically conclude that bread is nourishing. The conclusion can be verified experimentally. However, needless to say, the assumptions are absurd.

On the other hand, mathematics proves to be even sturdier than the physics it purportedly describes. Joseph Fourier (1768–1830) wrote a complete and elaborate mathematical theory of the conduction of heat that seemed to apply to the caloric theory that heat was a fluid, a theory that has long since been discarded. As Edmund Burke put it, "But too often different is rational conjecture from melancholic fact." However, Fourier's mathematics has proved to be essential in the analysis of musical sounds and other phenomena.

There is indeed some reason to question what mathematics tells us about what is real. Scientists wrestle with a problem, but the solution is not unique. In their efforts to build a theory they latch onto whatever mathematics may serve to build it. They make use of the available tools much as a man might use a hatchet in place of an ax and do a good enough job. In fact the whole history of physics tells us that newer theories replace older ones, just as relativity replaced Newtonian mechanics and quantum theory replaced the older atomic physics. Nor has relativity helped much thus far to explore the universe beyond our planetary system. Despite remarkable successes such as placing a man on the moon and photographing Saturn, we cannot claim the truth of mathematical physics.

Even time and space, unlike mass and force, cannot be perceived. Mass we experience as something substantial and force as the sensation of muscular exertion. However, time and space are constructs. We do have sensations of "thereness," position, size, and a sense of vastness. These are the sensory roots of space. Time also has some sensory basis in the succession of events. These fragmentary aspects of space and time are united by abstraction. Space and time may not be foundation stones on which we should build our knowledge of the real world. They are fundamental in relativity but may not be appropriate for quantum mechanics. Even length presupposes a rigid measuring rod, but it may not be rigid. Varying temperatures over some region may change its size without our being aware of it. Similar variability exists for area and volume.

Our mathematical theory of the physical world is not a description of the phenomena as we perceive them but a bold symbolic construction. Mathematics, released from the bondage of sensory experience, no longer describes reality but makes models of reality that serve the purposes of explanation, calculation, and prediction.

Whereas until around 1850, mathematical order and harmony were believed to be inherent in the design of the universe and mathematicians strove to uncover that design, the newer view, forced on mathematicians by their own creations, is that they are the legislators who decide what the laws of the universe should be. They impose whatever plan or order succeeds in describing restricted classes of phenomena that for inexplicable reasons continue to obey the laws. Does this last fact mean that there is an ultimate law and order that mathematicians approximate more and more successfully? There is no answer to this, but at the very least, faith in mathematical design had to be replaced by doubt. Yet what of the calamities of nature—earthquakes, meteorites striking the Earth, volcanoes, plagues—the unanswered questions of cosmogony, and our ignorance of what lies beyond our ken in our own galaxy, to say nothing of other problems facing humanity, do these not deny any likelihood of ultimate order? What we have achieved by way of mathematical description and prediction amounts to the good luck of the man who finds a hundred-dollar bill while casually taking a walk.

The history of physics is strewn with the wreckage of discarded theories. The recurrent hope that the complexities of nature will be circumscribed by some finite system of laws seems doomed to perpetual disappointment. It would be boldness indeed on our part to assume that these lessons of the past will not be projected into the future, and that our own current theories will survive the holocaust of time and experience. Our carefully erected systems are only more or less useful models of what we take temporarily to be the truth. None of the mathematical sciences is able to claim a unique, essential grasp of reality. It is not true that physics is objective whereas poetry and politics are not. All are concerned with truths, and none is more privileged than another. However, none rivals physical theory in precision and prediction. There is something in the external world that mathematical theory can capture and encapsulate.

We have a science of nature as humanity thinks about and describes it. Science stands between humanity and nature. But in the light of quantum theory the elementary particles are not real in the same sense as stones or trees but appear as abstractions derived from the real material of observation. If it does become impossible to attribute to the elementary particles existence in the truest sense, it becomes more difficult to consider matter as truly real.

Although Blaise Pascal (1623–1662) believed that the mathematical laws of nature were truths, he qualified the applicability of mathematics: "Correctness and truth are points so fine that our instruments are too blunt to touch them exactly. If the instruments get to the point, they hide it and cover likewise the adjacent space, thus resting more on the false than on the true."

Others go still further. P. W. Bridgman, in his *The Logic of Modern Physics* (1946), stated: "It is the merest truism evident at once to unsophisticated observation, that mathematics is a human invention." Clearly, then, our mathematics is subject to errors much as all humans are. Our achievement in physical theory reduces to a collection of mathematical relationships somewhat related to observable phenomena and to making predictions about physical phenomena some of which are not even observable at all, as in the case of electromagnetic waves. Abstract reasoning enables us to go beyond pictures drawn from sensations, although we are not entirely free of the latter.

These various explanations of the extent to which mathematics reflects or represents the truth about the physical world must be distinguished from many other statements that assert the truth of mathematics itself and its objective reality while they do not necessarily assert anything about the relationship to the external world. Thus Plato asserts in his dialogue *Menon* that mathematical structures are independent of experience and even precede it. For Plato, the existence of mathematics was actually a proof of the existence of an immortal soul, because the theorems could not be obtained from experience and so must accompany the soul into existence. The discovery of the theorems is really a recall of what is in one's memory.

Before 1800 or so, such views were held by all mathematicians, but some later mathematicians continue to maintain them. William R. Hamilton (1805–1865), although he invented the very objects (quaternions) that led to the questioning of the truth of arithmetic, maintained a position very much like Descartes's:

> Those purely mathematical sciences of algebra and geometry are sciences of the pure reason, deriving no weight and assistance from experiment, and isolated or at least isolable from all outward and accidental phenomena.... They are, however, ideas which seem so far born within us that the possession of them in any conceivable degree is only the development of our original powers, the unfolding of our proper humanity.

Arthur Cayley, one of the leading nineteenth-century algebraists, said in an address to the British Association for the Advancement of Science (1883) that "we are ... in possession of cognitions a priori, independent, not of this or that experience, but absolutely so of all experience. ... These cognitions are a contribution of the mind to the interpretation of experience."

Whereas men such as Hamilton and Cayley saw mathematics as embedded in the human mind, others see it existing in a world outside of man. That the belief in a unique objective world of mathematical truths independent of man should have been held before 1900 is certainly understandable. Even Gauss, who first appreciated the significance of non-Euclidean geometry, maintained the truth of number and analysis. The leading French mathematician of this century, Jacques Hadamard (1865–1963), affirmed in his *Psychology of Invention in the Mathematical Field:* "Although the truth is not yet known to us, it *pre-exists,* and inescapably imposes on us the path we must follow." And David Hilbert, at the International Congress in Bologna in 1928, asked: "How would it be with the truth of our knowledge and with the existence and progress of science if there were no truth in mathematics?"

Geoffrey H. Hardy (1877–1947), an outstanding analyst, expressed somewhat the same view in his book *A Mathematician's Apology:* "I believe that mathematical reality lies outside us, that our function is to discover or *observe* it, and that the theorems which we prove, and which we describe grandiloquently as our 'creations,' are simply our notes of our observations." Mathematicians merely discover concepts and their properties.

Some of these affirmations come from twentieth-century thinkers who were not much, if at all, concerned with the foundations of mathematics. What is surprising is that even some of the leaders in the work on foundations—David Hilbert, Alonzo Church, and the members of the Bourbaki school (see Chapter XII)—affirm that the mathematical concepts and properties exist in some objective sense and that they can be apprehended by human minds. Thus, mathematical truth is discovered not invented; what evolves is not mathematics but human knowledge of mathematics.

These assertions about the existence of an objective, unique body of mathematics do not explain where mathematics resides. They say merely that mathematics exists in some extrahuman world, a castle in the air, and is merely detected by humans. The axioms and theorems are not purely human creations; instead, they are like riches in a mine that have to be brought to the surface by patient digging. Yet their existence is as independent of man as the planets appear to be.

Is then mathematics a collection of diamonds hidden in the depths of the universe and gradually unearthed, or is it a collection of synthetic stones manufactured by humans, yet so brilliant nevertheless that they bedazzle those mathematicians who are already partially blinded by pride in their own creations?

The second view—that mathematics is entirely a product of human thought—is held by a school of mathematicians known as intuitionists and can be traced back to Aristotle. Whereas some assert that truth is guaranteed by the mind, others maintain that mathematics is the creation of fallible human minds rather than a fixed body of knowledge.

Hermann Hankel, Richard Dedekind, and Karl Weierstrass all believed that mathematics is a human creation. Dedekind affirmed in a letter to Heinrich Weber: "I advise moreover that we understand by number not the class itself, but something new . . . which the mind creates. We are of a divine race and we possess . . . the power to create." Weierstrass endorsed this thought with the words, "The true mathematician is a poet." Ludwig Wittgenstein, a student of Russell and an authority in his own right, also believed that the mathematician is an inventor not a discoverer. All these men and others conceive of mathematics as something far beyond bondage to empirical findings or rational deductions. In favor of their position is the fact that such elementary concepts as irrational numbers and negative numbers are neither deductions from empirical findings nor entities obviously existing in some external world.

Those who favor the view that mathematics is manmade are in essence Kantians, for Immanuel Kant (1724–1804) placed the source of mathematics in the organizing power of the mind. However, the modernists say that it is not in the morphology or physiology of the mind that mathematics originates; instead, it is in the activity of the mind. It organizes by methods that are evolutionary. The creative activity of the mind constantly evolves newer and higher forms of thought. In mathematics the human mind is able to see clearly that it is free to create a body of knowledge that it finds interesting or useful. Moreover, the field of creation is not a closed one. Notions that apply to existing and newly arising fields of thought will be created. The mind has the power to devise structures that will embrace the data of experience and provide a mode of arranging them. The source of mathematics is the progressive development of the mind itself.

The present conflicts about the nature of mathematics itself and the fact that mathematics is not today a universally accepted, indisputable body of knowledge certainly favor the view that mathematics is a human creation. As Einstein put it, "whoever undertakes to set himself up as a judge in the field of Truth and Knowledge is shipwrecked by the laughter of the gods."

Mathematicians had given up God; so it behooved them to accept man, and this is what they have done. They have continued the development of mathematics and the search for laws of nature, knowing that what they produced was not the design of God but the work of man. Their past successes helped them to retain confidence in what they were doing, and fortunately, hosts of new successes greeted their efforts. What has preserved the life of mathematics was the powerful medicine humans had themselves concocted—the enormous achievements in celestial mechanics, acoustics, hydrodynamics, optics, electromagnetic theory, and engineering, and the incredible accuracy of its predictions. Thus, mathematical creation and application to science have continued at an even faster pace.

All of this development Sir James Jeans sums up in his *The Mysterious Universe* (1930) as follows:

> Our remote ancestors tried to interpret nature in terms of anthropomorphic concepts of their own creation and failed. The efforts of our nearer ancestors to interpret nature on engineering lines proved equally inadequate. . . . On the other hand, our efforts to interpret nature in terms of the concepts of pure mathematics have, so far, proved brilliantly successful. It would now seem to be beyond dispute that in some way nature is more closely allied to the concepts of pure mathematics than to those of biology or of engineering.

In this latest outcome Jeans sees a close kinship between man and the physical universe, and "it can hardly be disputed that nature and our conscious mathematical minds work according to the same laws." Somewhat cautiously he adds that "the universe can best be pictured, although still very imperfectly and inadequately, as consisting of pure thought, the thought of what, for want of a wider word, we must describe as a mathematical thinker." Those who still deplore that to achieve success the physical sciences have to pay the price of mathematical abstractness must reconsider what it is they would look for in our ultimate scientific exposition of the nature of the physical world.

Despite or regardless of what the newer philosophical doctrines may say about the existence and our knowledge of the physical world, one fact seems indisputable. The new physics has turned away from mechanical models or even pictures of physical realities and has emphasized and even concentrated on mathematical description, a trend that, as far as one can foretell, will continue and from which there is no likelihood of any reversal. The newer realms of physics are so remote from ordinary experience, from sensuous perception, that only mathematics can grasp them.

As Jeans put it, "The making of models or pictures to explain mathematical formulae and the phenomena they describe, is not a step towards, but a step away from, reality; it is like making graven images of a spirit."

As in Plato's allegory of the cave wherein humans see only shadows of people and events, so we who live in the physical world see only shadows of many physical phenomena, and the shadows are mathematical. There may not be ghosts, witches, or devils, but there are physical phenomena as imperceptible and as intangible as any creations of the human imagination.

The trend toward accepting mathematical laws as the reality is evident in many writings. J. W. N. Sullivan, in his *The Limitations of Science* (1933), says that only quantitative aspects of material phenomena are concerned with the real world. In particular, the new science does not require that we know the nature of the entities we discuss but only their mathematical structure. In fact, Jeans, in his *The Mysterious Universe*, aptly describes the universe as a great thought. Mind is no longer an accidental intruder but the creator and governor of the world of matter.

Apropos of quantum mechanics in particular, the physicist and philosopher Professor Henry Margenau insists that the wave functions of Schrödinger are the true reality.

Perhaps the best summarizing statement of the position taken by most scientists today was made by Einstein in his *The World as I See It* (1934):

> Our experience hitherto justifies us in believing that nature is the realization of the simplest conceivable mathematical ideas. I am convinced that we can discover by purely mathematical constructions the concepts and the laws connecting them with each other, which furnish the key to the understanding of natural phenomena. . . . Experience remains, of course, the sole criterion of the physical utility of a mathematical construction. But the creative principle resides in mathematics. In a certain sense, therefore, I hold it is true that pure thought can grasp reality, as the ancients dreamed.

Endowed with a few limited senses and a brain, humans began to pierce the mystery about them. By utilizing what the senses reveal immediately or what can be inferred from experiments, humans adopted axioms and applied their reasoning powers. Their quest was the quest for order; their goal was to build systems of knowledge as opposed to transient sensations, and to form patterns of explanation that might help them attain some mastery over their environment. Their chief accomplishment, the product of human reason itself, is mathematics. It is not a perfect gem, and continued polishing will probably not remove all flaws. Nevertheless, mathematics has been our most effective link with the world of sense perceptions, and it is still the most precious jewel of the human mind that must be treasured and husbanded. It has been in the van of reason and no doubt will continue to be even if new flaws are discovered by more searching scrutiny.

The waves of mathematical thought are constantly pounding the rocky coasts, which bar their smooth, quiet entry onto the lands they seek to embrace. Still, the pounding of centuries has worn down even blocks of granite and has cleared the way for the engulfment of new areas.

XII

Why Does Mathematics Work?

The eternal mystery of the world is its comprehensibility.
Albert Einstein

Life is the art of drawing sufficient conclusions from insufficient premises.
Samuel Butler

In view of the conflicting views about the nature of mathematics and its relationship to the physical world, we must ask why mathematics works at all. We must face the fact that there is no universally accepted correspondence between mathematics and physical reality. Yet the many successful predictions of what is physically real—for example, electromagnetic waves, the predictions of relativity, the mathematical interpretation of what little is observable in atomic phenomena, and even the Newtonian theory of gravitation in its time, to say nothing of hundreds of successful predictions we have not surveyed—do call for some explanation.

Hence man faces a twofold mystery. Where the physical phenomenon is understood and we accept the relevant axioms, why do the hundreds of deductions from axioms prove to be as applicable as the axioms themselves? Does nature conform to man's logic? And, second, why does it work in domains where the physical phenomena are unknown? One cannot dismiss these questions lightheartedly. Far too much of our science and technology depends on mathematics. Surely, there must be some power or resource in mathematics that is not apparent.

In Greek times, where just one body of mathematics was conceived and the applications were far more limited, the answers given were by modern standards simplistic and rather dogmatic. Likewise, to the thinkers of the sixteenth, seventeenth, and eighteenth centuries the answer to the question of why mathematics works was straightforward. Fully impressed with the Greek belief that the world was designed mathematically, and equally impressed with the medieval doctrine that God designed the world mathematically, they regarded mathematics as the road to truths about nature. In other words, by making God a devoted and supreme mathematician, it became possible to view the search for the mathematical laws of nature as a religious quest. The study of nature became the study of God's word, His ways, and His will. The harmony of the world was God's mathematical arrangement. God put into the world that rigorous mathematical order that we comprehend only laboriously. Mathematical knowledge was absolute truth and as sacrosanct as any line of Scripture; it was superior in fact, for there is much disagreement about the Scriptures but there could be none about mathematical truths.

Thus, Catholic emphasis on a universe rationally designed by God and the Pythagorean–Platonic insistence on mathematics as the fundamental reality of the physical world were fused in a program for science that in essence amounted to this: science was to discover the mathematical relationships that underlie and explain all natural phenomena and thus reveal the grandeur and glory of God's handiwork. As John Herman Randall says in the *Making of the Modern Mind*, "Science was born of a faith in the mathematical interpretation of Nature, held long before it had been empirically verified."

Of the philosophers who urged mathematics on the modern world as the road to reality, René Descartes (1596–1650) was most influential. Although his approach was not the lasting one, he was the last of the scholastics as well as the first of the moderns in that his emphasis on mathematical reason was pronounced.

Descartes took up the question of how the mathematical creations of the human mind can be trusted to give us knowledge about the physical world. His answer, as we noted earlier, was to trust in God. Descartes believed that the mind has innate ideas of space, time, numbers, and God, and also recognizes other intuitions as truths. This knowledge is indubitable. Thus, the idea of God could not come from sensations, because eternity, omniscience, omnipotence, and perfection are not evident in the physical world. The mind also has the idea of an external world. Is there one? God would not deceive us. On the other hand, for Descartes sense perceptions are sense deceptions. Fortunately from the truths of mathematics recognized by the mind independently of experience, one could use reasoning to deduce new truths about the physical world. How could we be sure that the reasoning was correct? Again Descartes called on God: He *causes* our reasoning to agree with reality.

Descartes's belief in the mathematical design of nature was affirmed by his contemporaries and by his successors of the next couple of centuries. Kepler, too, affirmed that the reality of the world consists of its mathematical relations. Mathematical principles are the alphabet in which God wrote the world, said Galileo; without their help it is impossible to comprehend a single word, and humanity wanders in vain through dark labyrinths. In fact, only the mathematically expressible properties of the physical world are really knowable. The universe is mathematical in structure and behavior, and nature acts in accordance with inexorable and immutable laws. In one of his letters, Galileo goes so far as to say: "Yet for my part any discussion of the Sacred Scriptures might have lain dormant forever; no astronomer or scientist who remained within proper bounds has ever gotten into such things." Of course, Galileo believed in God's mathematical design, and all he meant by this statement is that no mystical or supernatural forces should be invoked to explain the workings of nature.

Newton, too, believed that God had designed the world according to mathematical principles. In a letter to Richard Bentley of December 10, 1692, Newton said, "When I wrote my treatise about our system [*Mathematical Principles of Natural Philosophy*] I had an eye on such principles as might work with considering men for the belief in a Deity; and nothing can rejoice me more than to find it useful for that purpose."

Newton regarded the chief value of his scientific work to be the support of revealed religion. He was a learned theologian, although he never took orders. He thought scientific work hard and dreary, but he stuck to it because it gave evidence of God's work. Like his predecessor Isaac Barrow, Newton turned to religious studies later in life. He still believed in a designed universe, but he banked on God to keep the world functioning according to plan. He used the analogy of a watchmaker keeping a watch in repair.

Although Gottfried Wilhelm Leibniz (1646–1716) was a universal man and his contributions to mathematics, notably to the calculus, were of the first order, he did not extend the reign of mathematics to any broad branch of science. More like Descartes, his philosphy of science was his most influential contribution to the doctrine of the mathematical design of nature.

In his *Essais de Théodicée* Leibniz affirmed the familiar thought that God is the intelligence who created this carefully designed world.

His explanation of the concord between the real and mathematical worlds and the ultimate defense of the applicability of his calculus to the real world were the unity of the world and God. *Cum Deus calculat, fit mundus* (As God calculates, so the world is made). Sense and reason derived from God. The laws of reality cannot therefore deviate from the ideal laws of mathematics. The universe was the most perfect conceivable, the best of all possible worlds, and rational thinking disclosed its laws.

What Descartes, Kepler, Galileo, Newton, Leibniz, and many other founders of modern mathematics believed can be expressed thus: there is inherent in nature a hidden harmony that reflects itself in our minds in the form of simple mathematical laws. This harmony is the reason that events in nature are predictable by a combination of observation and mathematical analysis. This presupposition, even in earlier times, found fulfillments beyond expectation.

Of course, the mathematical design of nature had to be uncovered by humanity's continual search. God's ways appeared to be mysterious, but it was certain that they were mathematical and that human reasoning would in time discern more and more of the rational pattern that God had utilized in creating the universe. That humanity reasoned exactly in the way in which God had planned seemed readily understandable on the ground that there could be but one brand of correct reasoning.

William James, in his essay *Pragmatism* (1907), describes the attitude of the mathematicians of this period in this way:

> When the first mathematical, logical and natural uniformities, the first *laws,* were discovered, men were so carried away by the clearness, beauty and simplification that resulted that they believed themselves to have deciphered authentically the eternal thoughts of the Almighty. His mind also thundered and reverberated in syllogisms. He also thought in conic sections, squares and roots and ratios, and geometrized like Euclid. He made Kepler's laws for the planets to follow; he made velocity increase proportionally to the time in falling bodies, he made the law of sines for light to obey when refracted; . . . He thought the archetypes of all things and devised their variations; and when we rediscover any one of these wondrous institutions, we seize his mind in its very literal intention.

From a historical standpoint it is ironic that God's role became more and more neglected as universal laws embracing heavenly and earthly motions began to dominate the intellectual scene, and as the continued agreement between predictions and observations bespoke the perfection of the laws. God sank into the background, and the mathematical laws of the universe became the focus of attention. Leibniz saw some of the implications of Newton's *Mathematical Principles,* a world functioning according to plan with or without God, and attacked the book as anti-Christian.

The concern for the attainment of purely mathematical results gradually replaced the homage to God's design. Although many a mathematician continued to believe in God's presence, His design of the world, and mathematics as the science whose main function was to provide the tools to decipher God's design, His presence became dimmer during the latter half of the eighteenth century. The further the development of mathematics proceeded in the eighteenth century and the more numerous its successes, the more the religious inspiration for mathematical work receded.

The gradual elimination of God in the mathematical study of nature changed from orthodoxy to various intermediate stages such as rationalistic supernaturalism, Deism, agnosticism, and outright atheism. These movements had their effect on the mathematicians who, in the eighteenth century, were cultured people. As Denis Diderot (1713–1784), certainly one of the leading intellectuals of his age, put it, "If you want me to believe in God, you must make me touch him." Augustin-Louis Cauchy (1789–1857), a devout Catholic, said humanity "rejects without hesitation any hypothesis that is in contradiction with revealed truth"; nevertheless, the belief in God as the designer of the universe practically vanished. As the renowned mathematician Jean Le Rond d'Alembert (1717–1783), chief collaborator of Denis Diderot in the writing of the famous French *Encyclopédie,* put it, "The true system of the world has been recognized, developed and perfected." Natural law clearly was mathematical law.

Lagrange and Laplace, though of Catholic parentage, were agnostics. In fact, Laplace completely rejected any metaphysical principles that were based on an existent God. There is a well-known story that when Laplace gave Napoleon a copy of his *Mécanique céleste,* Napoleon remarked, "Monsieur Laplace, they tell me you have written this large book on the system of the universe and have never even mentioned the Creator." Laplace is said to have replied, "I have no need of this hypothesis." Nature replaced God. The mathematicians plunged ahead in their search for the mathematical laws of nature as though hypnotized into believing that they, the mathematicians, were the anointed ones to discover what had been attributed to God's design.

At any rate, by the end of the eighteenth century mathematics was like a tree firmly grounded in reality, with roots already two thousand years old, with majestic branches, and towering over all other bodies of knowledge. Surely such a tree would live forever. The conviction that nature was mathematically designed was firmly held. To uncover that design and understand the laws by which the universe was regulated was the task of the mathematician, and mathematics proper was the tool for the task. Through industry, persistence, and intense labor more truths could be obtained.

The development of non-Euclidean geometry (Chapter VIII) showed that humanity's mathematics did not speak for nature, much less lead to a proof of the existence of God. It became clearer, too, that humanity institutes the order in nature, the apparent simplicity, and the mathematical pattern. Nature itself might have no inherent design, and perhaps the best one could say of mathematics was that it offered no more than a limited, workable, rational plan.

The nineteenth century brought with it more humble goals. Evariste Galois (1811–1832) said of mathematics, "This science is the work of the human mind, which is destined rather to study than to know, to seek the truth rather than to find it." Perhaps it is in the nature of truth that it wishes to be elusive. Or, as the Roman philosopher Lucius Seneca (c. 4 B.C.–A.D. 65) put it, "So, too, Nature does not at once disclose all Her mysteries."

Still, if mathematics lost its place in the citadel of truth, it remained at home in the physical world. The major, inescapable fact, and the one that still has inestimable importance, is that mathematics is the method par excellence by which to investigate, discover, and represent physical phenomena. In some branches of physics it is, as we have seen, the essence of our knowledge of the physical world. If mathematical structures are not in themselves the reality of the physical world, they are the only key we possess to that reality. The creation of non-Euclidean geometry not only did not destroy this value of mathematics or confidence in its results, but—rather paradoxically—increased its usefulness, because mathematicians felt freer to investigate radically new ideas, finding some of these to be applicable. In fact, the role of mathematics in the organization and mastery of nature has expanded at an almost incredible rate since 1830. In addition, the accuracy with which mathematicians can represent and predict natural occurrences has increased remarkably since Newton's day.

We seem, therefore, to be confronted with a paradox. A subject that no longer lays claim to truth has furnished the marvelously adaptable Euclidean geometry, the pattern of the extraordinarily accurate heliocentric theory of Copernicus and Kepler, the grand and embracing mechanics of Galileo, Newton, Lagrange, and Laplace, the physically inexplicable but widely applicable electromagnetic theory of Maxwell, the sophisticated theory of relativity of Einstein, and much of atomic structure. All of these highly successful developments rest on mathematical ideas and mathematical reasoning. Is there perhaps some magical power in the subject that, although it had fought under the invincible banner of truth, has actually achieved its victories through some inner mysterious strength?

The problem has been posed repeatedly, notably by Albert Einstein in his *Sidelights on Relativity*:

> Here arises a puzzle that has disturbed scientists of all periods. How is it possible that mathematics, a product of human thought that is independent of experience, fits so excellently the objects of physical reality? Can human reason without experience discover by pure thinking properties of real things?

Although Einstein understood that the axioms of mathematics and principles of logic derive from experience, he was questioning why the many deductions from these axioms and principles, which are made by human minds, should still fit experience.

The question of why mathematics works has been answered in various ways. One answer is that mathematicians change the axioms to make the resulting deductions fit experience. This thought was first expressed by Diderot in his *Pensées sur l'interprétation de la nature* (1753). He said the mathematician was like a gambler; both played games with abstract rules that they themselves had created. Their subjects of study were only matters of convention that had no foundation in reality. Equally critical was the intellectual Bernard Le Bovier de Fontenelle (1657–1757). He attacked the belief in the immutability of the laws of the heavenly motions with the observation that as long as roses could remember no gardener ever died.

This is the position taken by modern model builders. One starts with a possible model, deduces consequences, and then checks against experience. If the model is not adequate, one can change it. Nevertheless, the deduction of hundreds of applicable theorems from any one model still raises a question that is not readily answered.

Another modern and quite different explanation that stems from Kant is still offered, albeit perhaps in modified form. Kant maintained that we do not and cannot know nature; rather, we have sense perceptions, and our minds, endowed with established structures (intuitions was Kant's term) about space and time, organize these perceptions in accordance with what these built-in mental structures dictate. Thus, we organize spatial perceptions in accordance with the laws of Euclidean geometry because our minds require this. Having so organized spatial perceptions, they continue to obey the laws of Euclidean geometry. (Of course, Kant was wrong in insisting on Euclidean geometry.) In other words, we see only what our mathematical "optics" permits us to see. "The understanding," Kant wrote, "does not draw its laws from nature but prescribes them to nature."

The physicist Arnold Sommerfeld (1868–1951), in common with many other scientists, believed that there is an intolerable arrogance in the idea of prescribing to nature. However, Sir Arthur Stanley Eddington (1882–1944) supported the Kantian view:

> [We] have found that where science has progressed the farthest, the mind has but regained from nature that which the mind has put into nature.

> We have found a strange footprint on the shores of the unknown. We have devised profound theories, one after another, to account for its origin. At last, we have succeeded in reconstructing the creature that made the footprint. And Lo! it is our own.

Eddington came to believe that the universe of human experience is essentially the creation of the human mind and that if we could only understand how the mind works, we could derive the whole of physics—presumably all of science as well—by purely theoretical methods, barring certain dimensional constants that are matters of accident depending on where one happens to be in the universe.

Jules Henri Poincaré (1854–1912) offered another explanation that is in large measure Kantian, although it is now called conventionalism. In his *Science and Hypothesis* he says:

> Can we maintain that certain phenomena which are possible in Euclidean space would be impossible in non-Euclidean space, so that experiment in establishing these phenomena would directly contradict the non-Euclidean hypothesis; I think that such a question cannot seriously be asked. Experiment plays a considerable role in the genesis of geometry; but it would be a mistake to conclude from that that geometry is, even in part, an experimental science. If it were experimental, it would only be approximate and provisory. And what a rough approximation it would be! Geometry would be only the study of the movements of solid bodies; but in reality, it is not concerned with natural solids; its object is certain ideal solids, absolutely invariable, which are but a greatly simplified and very

remote image of them. The concept of these ideal bodies is entirely mental, and experiment is but the opportunity which enables us to reach the idea. . . .

Experiment guides us in this choice, which it does not impose on us. It tells us not what is the truest, but what is the most convenient geometry. I challenge anyone to give me a concrete experiment which can be interpreted in the Euclidean system, and which cannot be interpreted in the system of Lobatchevsky. As I am well aware that this challenge will never be accepted, I may conclude that no experiment will ever be in contradiction with Euclid's postulate; but, on the other hand, no experiment will ever be in contradiction with Lobatchevsky's postulate.

Poincaré believed that there are an infinite number of theories that can adequately explain and describe every branch of experience. The choice of theory is arbitrary, although simplicity is a good guide. We invent and use ideas that seem to work; other theories might work if sufficient effort is put into them. Although Poincaré was more explicit in explaining how mathematics is made to work, he did agree somewhat with the Kantian explanation in that he believed the accord between mathematics and nature is fashioned by human minds. In *The Value of Science,* he affirmed:

Does the harmony which human intelligence thinks it discovers in Nature exist apart from such intelligence? Assuredly no. A reality completely independent of the spirit that conceives it, sees it or feels it, is an impossibility. A world so external as that, even if it existed, would be forever inaccessible to us. What we call "objective reality" is, strictly speaking, that which is common to several thinking beings and might be common to all; this common part, we shall see, can only be the harmony expressed by mathematical laws.

The philosopher William James expressed the same idea in his *Pragmatism:* "All the magnificent achievements of mathematical and physical science . . . proceed from our indomitable desire to cast the world into a more rational shape in our minds than the shape into which it is thrown there by the crude order of our experience."

In his *Aspects of Science* (second series), J. W. N. Sullivan put the point even more strongly: "We are the law-givers of the universe; it is even possible that we can experience nothing but what we have created and that the greatest of our mathematical creations is the material universe itself."

These men assert that scientific truth is made, not found. Even if the implications are experimentally verifiable, they are merely symptoms of physical truth.

Einstein in 1938 essentially supported the Kantian view:

> Physical concepts are free creations of the human mind, and are not, however it may seem, uniquely determined by the external world. In our endeavor to understand reality we are somewhat like a man trying to understand the mechanism of a closed watch. He sees the face and the moving hands, even hears it ticking, but he has no way of opening the case. If he is ingenious he may form some picture of the mechanism which could be responsible for all the things he observes, but he may never be quite sure his picture is the only one which could explain his observations. He will never be able to compare his picture with the real mechanism and he cannot even imagine the possibility of the meaning of such a comparison.

Einstein did believe that humanity's mathematics is at least partially governed by reality. In *The Meaning of Relativity* (1945) he said:

> For even if it should appear that the universe of ideas cannot be deduced from experience by logical means, but is, in a sense, a creation of the human mind, without which no science is possible, nevertheless this universe of ideas is just as little independent of the nature of our experiences as clothes are of the form of the human body.

Still another explanation of why mathematics works is essentially a reversion to the seventeenth- and eighteenth-century belief that the world is mathematically designed, although in modern times the associated religious beliefs of those earlier centuries have been abandoned. This is the position that Sir James Jeans (1877–1946), one of the greatest physicists of our age, expressed in his *The Mysterious Universe:*

> The essential fact is simply that all the pictures which science now draws of nature, and which alone seem capable of according with observational fact, are *mathematical* pictures. . . . Nature seems very conversant with the rules of pure mathematics. . . . In any event it can hardly be disputed that nature and our conscious mathematical minds work according to the same laws.

Like Jeans, Pierre Duhem, one of the great historians and philosophers of science, in his *The Aim and Structure of Physical Theory,* passes through doubts to positive affirmation. He first describes a physical theory as "an abstract system whose aim is to summarize and classify logically a group of experimental laws without claiming to explain these laws." Theories are approximate, provisional, and "stripped of all objective references." Science is acquainted only with sensible appearances, and we should shed the illusion that in theorizing we are "tearing the veil from these sensible appearances." Furthermore, when a scientist of genius brings mathematical order and clarity into the confusion of appearances, he achieves his aim only at the expense of replacing relatively intelligible concepts by symbolic abstractions that reveal nothing of the true nature of the universe. Still, Duhem ends by declaring that "it is impossible for us to believe that this order and this organization produced by theory are not the reflected image of a real order and organization."

The fine nineteenth-century analyst Charles Hermite (1822–1901) expressed his belief that there is an objective real world described by mathematics. He said in a letter to the mathematician Stieltjes:

> I believe that the numbers and functions of analysis are not the arbitrary product of our spirits: I believe that they exist outside of us with the same character of necessity as the objects of objective reality; and we find or discover them and study them as do the physicists, chemists and zoologists.

On another occasion he said, "We are servants rather than masters in mathematics."

Hermann Weyl, in his *Philosophy of Mathematics and Natural Science* (1949), said:

> There is inherent in nature a hidden harmony that reflects itself in our minds under the image of simple mathematical laws. That then is the reason why events in nature are predictable by a combination of observation and mathematical analysis. Again and again in the history of physics this conviction, or should I say this dream, of harmony in nature has found fulfillments beyond our expectation.

Perhaps, however, the wish was father to the thought, for in his book he adds: "And yet science would perish without a supporting transcendental faith in truth and reality, and without the continuous interplay between its facts and constructions on the one hand and the imagery of ideas on the other."

More surprising is that Weyl also agreed that soundness may be judged by application to the physical world. Weyl had contributed much to mathematical physics, and he was not willing to sacrifice useful results. In his *Philosophy of Mathematics and Natural Science* he conceded:

> How much more convincing and closer to facts are the heuristic arguments and the subsequent systematic constructions in Einstein's general relativity theory, or the Heisenberg–Schrödinger quantum mechanics. A

truly realistic mathematics should be conceived, in line with physics, as a branch of the theoretical construction of the one real world, and should adopt the same sober and cautious attitude toward hypothetic extensions of its foundations as is exhibited by physics.

Here Weyl is certainly advocating treating mathematics as one of the sciences. Its theorems, like those of physics, may be tentative and precarious. They may have to be recast, but correspondence with reality is one sure test of soundness.

Another school of thought that one may characterize as empirical advocates that mathematics derives only approximately accurate laws to describe our knowledge of nature. Such an empirical foundation and test for mathematics was advocated by John Stuart Mill (1806–1873). He admitted that mathematics is more general than the several physical sciences. However, what "justifies" mathematics is that its propositions have been tested and confirmed to a greater extent than those of the physical sciences. Hence people came to think incorrectly of mathematical theorems as qualitatively different from confirmed hypotheses and theories of other branches of science. The theorems were taken as certain, whereas physical theories were thought of as very probable or merely corroborated by experience. Mill based his assertions on philosophical grounds. With all the more reason, many recent and current workers in the foundations have become pragmatic.

Andrzej Mostowski, one of the prominent and active workers in the foundations, agrees. At a congress held in Poland in 1953, he stated:

The only consistent point of view, which is in accord not only with healthy human understanding but also with mathematical tradition, is rather the assumption that the source and last raison d'être of the number concept—not only the natural but also the real numbers—lie in experience and practical applicability. The same is true of the concepts of set theory insofar as they are needed in the classical domains of mathematics.

Mostowski goes further. He says that mathematics is a natural science. Its concepts and methods have their origin in experience, and any attempts to found mathematics without regard to its origin in the natural sciences, its applications, and even its history are doomed to fail.

Willard Van Orman Quine, a currently active logicist, has also been willing to settle for physical soundness. In a 1958 article, part of the collection entitled *The Philosophical Bearing of Modern Logic,* he said:

We may more reasonably view set theory, and mathematics generally, in much the way in which we view theoretical portions of the natural sciences themselves; as comprising truths or hypotheses which are to be vindicated less by the light of pure reason than by the indirect systematic contribution which they make to the organizing of empirical data in the natural sciences.

Even Bertrand Russell, who in 1901 claimed that the edifice of mathematical truth, logical and physical, remained unshakable, admitted in an essay of 1914 that "our knowledge of physical geometry is synthetic, but not a priori." It is not deducible from logic alone. In the second edition of his *Principia* (1926) he conceded still more. Logic and mathematics, like Maxwell's equations of electromagnetic theory, "are believed because of the observed truth of certain of their logical consequences."

What these leaders are acknowledging is that mathematics is a human activity and is subject to all the foibles and frailties of humans. Any formal, logical account is a pseudo-mathematics, a fiction, even a legend, despite the element of reason.

The physicists, too, believe that mathematics is no more than the abstract, and only approximate, formulation of experience. The Nobel prize-winning physicist P. W. Bridgman says in his *The Nature of Physical Theory* (1936): "Mathematics thus appears to be ultimately just as truly an empirical science as physics or chemistry." Bridgman had no doubts. Theoretical science is a game of mathematical make-believe.

Ludwig Wittgenstein, one of the most profound philosophers of the subject, declared that mathematics is not only a human creation but is very much influenced by the cultures in which it was developed. Its "truths" are as dependent on human beings as is the perception of color or the English language.

Thus physicists (and some philosophers) believe that mathematics remains rooted in physical reality, and they invoke mathematics as an aid. For Planck, Mach, Boltzmann, and Helmholtz mathematics provides no more than a logical structure for the laws of physics.

A rather realistic account of the successes of mathematics vis-à-vis physical reality was made by Gilbert Lewis in his *The Anatomy of Science* (1926):

The scientist is a practical man and his are practical aims. He does not seek the *ultimate* but the *proximate*. He does not speak of the last analysis but rather of the next approximation. His are not those beautiful structures so delicately designed that a single flaw may cause the collapse of the whole. The scientist builds slowly and with a gross but solid kind of masonry. If dissatisfied with any of his work, even if it be near the very foundations, he can replace that part without damage to the remainder. On the whole, he is satisfied with his work, for while science may never be wholly right it certainly is never wholly wrong; and it seems to be improving from decade to decade.

The theory that there is an ultimate truth, although very generally held by mankind, does not seem useful to science except in the sense of a horizon toward which we may proceed, rather than a point which may be reached.

The position taken by the physicists should serve to remind us how much of our actual mathematics has developed out of our constant intercourse with the physical world around us. As William Barrett points out in his *The Illusion of Technique* (1978), the whole history of mathematics attests to this relationship between the mathematical mind and nature. Geometry and the calculus, for example, developed out of our need to deal with objects and phenomena of the physical world. Some modern mathematicians have tended to sever the tie with nature. In the exuberance of formalism they were led to imagine mathematics itself as a free excursion in the void. Modern philosophers have abetted this tendency. Admittedly, it is entirely unlikely that we should have been able to build airplanes or launch rockets without the aid of mathematics. The mistake was to take this or that isolated proposition and ask to what particular fact in the world it corresponded; and, of course, the answer would be negative. We do not isolate the single mathematical proposition from the body of mathematical discourse, and in turn we take this discourse as part of our total language; mathematics, as this functioning part, serves to tell us a great deal about the things of our world.

Moreover, Barrett says, it is precisely here that we may find the key to the question of conventionalism. The conventions we adopt must somehow "work"; that is, they must serve us in coping with nature. We might, for example, decide to change our mathematical conventions and drop the notion of irrational numbers altogether. It is this need to deal with nature that ultimately takes the measure of our various conventions—mathematical and others.

We need the notion of the mind as itself a product of nature and related to nature in its most fundamental modes of operation. The entities of mathematics do not subsist in a timeless Platonic world;

they are human constructs—but they are constructs that have their use and their being in relation to the natural world that encompasses them. All human thinking takes place against this background of nature. This view was neatly expressed by Alexander Pope:

> First follow nature and your judgment frame
> By her just standard, which is still the same.
> Unerring Nature, still divinely bright,
> One clear, unchanged and universal light,
> Life, force, and beauty must to all impart,
> At once the source, and end, the test of Art.
>
> . . .
>
> Those rules of old discovered, not devised,
> Are nature still, but nature methodized;
>
> . . .
>
> 'Tis Nature's voice, and Nature we obey.

Many mathematicians are happy to accept the remarkable applicability of mathematics but confess that they are unable to explain it. The distinguished group of mathematicians who write under the pseudonym Nicolas Bourbaki say there is an intimate connection between experimental phenomena and mathematical structures. Yet we are completely ignorant about the underlying reasons for this, and we shall perhaps always remain ignorant of them. In previous years mathematics was derived from experimental truths, especially from immediate space intuitions. However, quantum physics has shown that this macroscopic intuition of reality covers microscopic phenomena of a totally different nature, connecting it with fields of mathematics that had certainly not been thought of for the purpose of applications to experimental science. Hence there is nothing more than a fortuitous contact of two disciplines whose real connections are more deeply hidden than could have been supposed a priori. We can think of mathematics as a storehouse of mathematical structures, and certain aspects of physical or empirical reality fit into these structures, as if through a kind of preadaptation.

Charles Hermite expressed in a letter to Leo Koenigsberger (1837–1921) this same inability to explain the connection between mathematics and reality:

> These notions of analysis have their existence apart from us—they constitute a whole of which only a part is revealed to us, incontestably

although mysteriously associated with that other totality of things which we perceive by way of the senses.

Other thinkers have also felt obliged to admit that the marvelous power of mathematics remains inexplicable. The philosopher Charles Sanders Peirce (1839–1914) remarked, "It is probable that there is some secret here which remains to be discovered." More recently, Erwin Schrödinger in *What Is Life?* said that the miracle of humanity's discovering laws of nature may well be beyond human understanding. Another outstanding physicist, Freeman Dyson, agrees: "We are probably not close yet to understanding the relation between the physical and the mathematical worlds." To this one can add Einstein's remark: "The most incomprehensible thing about the world is that it is comprehensible." And yet Sir James Jeans affirmed that physical concepts and mechanisms are conjectured to construct the mathematical account, but then, paradoxically, that the physical aids are hardly more than fantasies; for Jeans the mathematical equations remain the only sure hold on the phenomena. The final harvest in physics will always be a collection of mathematical formulas; the real essence of material substance is forever unknowable.

All in all, the role of mathematics in modern science is now seen to be far more than that of a useful tool. The role has often been described as one of summarizing and systematizing in symbols and formulas what is physically observed or physically established by experimentation and then deducing from the formulas additional information that is accessible neither to observation, nor experimentation, nor information more readily obtained. However, this account of the role of mathematics falls far short of what it achieves. Mathematics is the essence of scientific theories, and the applications made in the nineteenth and twentieth centuries on the basis of purely mathematical constructs are even more powerful and marvelous than those made earlier when mathematicians operated with concepts suggested directly by physical phenomena. Although credit for the achievements of modern science—radio, television, airplanes, the telephone, the telegraph, high-fidelity phonographs and recording instruments, X rays, transistors, atomic power (and bombs), to mention a few that are familiar—cannot be accorded only to mathematics, the role of mathematics is more fundamental and less dispensable than any contribution of experimental science.

Whether or not these explanations of why mathematics works are acceptable, there is some justification for characterizing the new physics as mathematical rather than mechanical. Although Maxwell tried to invent a mechanical ether model while developing his electromagnetic theory, the completed structure is essentially mathematical; the "physical reality" to which the equations relate is a vague, nonmaterial concept of electromagnetic fields. Even Newton built his laws of mechanics as a purely mathematical structure.

It may be, as Eddington has said, that a knowledge of mathematical relations and structure is all that the science of physics can give us. And Jeans added that the mathematical description of the universe is the ultimate reality. The pictures and models (a very fashionable word today) we use to assist our understanding are a step away from reality. We go beyond the mathematical formulas at our own risk.

Because mathematics is a human creation, and because through mathematics we discover totally new physical phenomena, human beings create parts of their universe, gravity, electromagnetic waves, quanta of energy, and so forth. Of course, perceptions and experimentation give leads to the mathematician. There is a substratum of physical fact, but even where there is some physical reality, the full organization, completion, correction, and understanding come through mathematics.

What we know involves the human mind at least as much as what exists in the external world and even in the perceptions the human mind enters. To perceive a tree without recognizing the "treeness" is meaningless. Moreover, a collection of perceptions per se is meaningless. Humans and their minds are part of reality. Science can no longer confront nature as objective and humanity as the describer. They cannot be separated.

The dividing line between mathematical knowledge and empirical knowledge is not absolute. We constantly adjust the records of our observations and at the same time adjust our theories to meet new observations and experimental results. The objective in both efforts is a comprehensive and coherent account of the physical world. Mathematics mediates between man and nature, between man's inner and outer worlds.

We come finally to the undeniable and irresistible conclusion that our mathematics and physical reality are inseparable. Mathematics, insofar as it tells us what the physical world contains and insofar as the expression of that knowledge can only be in mathematical language and concepts, is as real as tables and chairs. There are boundaries to our knowledge of reality, but these are gradually pushed back.

It may be that humanity has introduced some limited and even artificial concepts and only in this way has managed to institute some order in nature. Our mathematics may be no more than a workable scheme. Nature itself may be far more complex or have no inherent design. Nevertheless, mathematics remains the method par excellence for the investigation, representation, and mastery of nature. In some domains it is all we have; if it is not reality itself, it is the closest to reality we can get.

Although it is a purely human creation, the access it has given us to some domains of nature enables us to progress far beyond all expectations. Indeed, it is paradoxical that abstractions so remote from reality should achieve so much. As artificial as the mathematical account may be, a fairy tale perhaps, it is one with a moral. To thoughtful scientists it has been a constant source of wonder that nature shows such a large measure of correlation with their mathematical formulas. Whether or not the uniformities expressed by scientific laws reside in nature and are discovered, or whether they are invented and applied to nature by the mind of the scientist, scientists in a spirit of humility should hope that through unremitting labors they might achieve a greater comprehension of nature's wonders.

XIII

Mathematics and Nature's Behavior

> *Our experience hitherto justifies us in believing that nature is the realization of the simplest conceivable mathematical ideas.*
>
> Albert Einstein

Science has determined our attitude toward nature ever since Greek times, but all the more so since major scientific theories have been confirmed in predictions hundreds and thousands of times. Major philosophies have been built on the existence and seemingly incontrovertible findings of physical science.

The more recent developments, particularly electromagnetic theory, relativity, and quantum theory, have forced reconsideration of philosophical doctrines. This chapter will provide sketches of and will contrast some of the older and newer doctrines shaping our views about nature. The mentality of an epoch and the thoughts and actions of society spring from the view of the world that is dominant. Dominant today is our view of the physical world.

A major doctrine vital in itself and, as we shall see, supportive of other major doctrines, has been called mechanism, sometimes described as materialism. Loosely stated for the moment, mechanism maintains that the physical world at least is a huge machine whose parts interact with one another. The machine acts flawlessly and infallibly. Witness the motions of the planets and the tides and the predictability of eclipses. The parts of the machine are pieces of matter in motion, which in turn is caused by the action of forces. Let us examine these concepts more closely.

Basic to mechanism is matter. The belief in matter as the essence of physical reality dates back to Greek times. Major Greek philosophers looked about themselves and studied nature as best they could with their very limited resources. However, they were prone to take off quickly from a few observations to affirm sweeping metaphysical generalities. Thus, Leucippus and Democritus advocated the idea of a universe consisting of indestructible and indivisible atoms existing in a void. Aristotle constructed matter from the "four elements," not of actual earth, water, air, and fire, but of four underlying entities having properties perceptible in these four realities.

Thomas Hobbes (1588–1679), in a somewhat crude form of this doctrine, asserted:

> The universe, that is, the whole mass of all things that are, is corporeal, that is to say, body, and hath the dimensions of magnitude, namely, length, breadth, and depth; also, every part of body is likewise body, and hath the like dimensions, and consequently every part of the universe is body, and that which is not body is no part of the universe; and because the universe is all, that which is no part of it is nothing, and consequently nowhere.

Body, he continued, is something that occupies space, is divisible and movable, and behaves mathematically.

Mechanism, then, may be said to assert that reality is merely a complex machine driving objects in space and time. Because we ourselves are part of physical nature, all of humanity must be explainable in terms of matter, motion, and mathematics.

Descartes, too (as we have previously noted), insisted that all physical phenomena could be explained in terms of matter and motion. Moreover, matter acted on matter by direct contact. Matter was composed of small invisible particles that differed in size, shape, and other properties. Because these particles were too small to be visible, it was necessary to make hypotheses about how these particles behaved to account for the larger phenomena that we can observe such as the motions of the planets around the sun. Descartes did not accept empty space. A vase whose interior was completely void of matter would, he believed, collapse.

However, Descartes's science, adopted by most pre-Newtonians, especially Huygens, proposed essentially the same function of science, namely, to provide a *physical* explanation of the action of natural phenomena.

The belief in matter as the essence of physical reality was maintained by all scientists and philosophers until about 1900. Newton says in his *Opticks:*

It seems very probable to me that God in the beginning formed matter in solid, massy, hard, impenetrable, movable particles, so very hard as never to wear and break into pieces, no ordinary power being able to divide what God himself made One in the first creation.

Because matter in motion was the key to a mathematical description of falling bodies and planetary motion, the scientists themselves attempted to fit such a materialistic explanation to phenomena whose nature they did not understand at all. Heat, light, electricity, and magnetism were regarded as imponderable kinds of matter, imponderable meaning merely that the densities of these kinds of matter were too small to be measured. The matter in heat, for example, was called caloric. A body when it was heated soaked up this matter, just as a sponge soaks up water. Electricity was, similarly, matter in the state of a fluid or two fluids, and these fluids flowing through wires were the electric current.

Matter was set into motion and generally kept in motion by the action of forces. A billiard ball hitting another billiard ball imparts motion by the force of impact. To account for the continuing motion of the planets, Newton introduced in particular the force of gravitation. To account for electric and magnetic phenomena, Faraday introduced lines of electric force and magnetic force that he believed were real.

Of the three concepts—matter, force, and motion—force acted on matter, and motion was a behavior of matter; hence matter was fundamental. The philosophers thereupon proclaimed matter that behaved in accordance with fixed mathematical laws as the sole reality.

The branch of physical science most fully developed by the end of the eighteenth century was mechanics. In the famous French *Encyclopédie,* d'Alembert and Diderot stated, rather overconfidently, that mechanics is the universal science. As Diderot put it, "The true system of the world has been recognized, developed and perfected." Mechanics did become the paradigm of newer and rapidly burgeoning branches of physical science.

Although Leibniz defended mechanism as a self-evident truth, he was not satisfied with mechanism alone. God, energy, and purpose were equally important to him. In his *Monadology* (1714), he affirmed that the universe was composed of little monads, each indivisible and a center of energy. Each contained its past and future. Monads worked together in a preconceived harmony to form larger organisms. They were the inner dynamism of things. Mechanism, however, treated only the external, spatial and other physical features such as force.

The master physicist, physician, and mathematician Hermann von Helmholtz (1821–1894) declared in an address reproduced in his *Popular Lectures on Science* (1869) that the final aim of all natural science is to resolve itself into mechanics. Helmholtz did recognize that not all the elements of mechanics were as yet understood, and he called specific attention to the problem of the nature of forces:

> Finally, therefore, we discover that the problem of physical material science is to refer natural phenomena back to unchangeable attractive and repulsive forces between bodies whose intensity depends wholly upon distance. The solubility of this problem is the condition for the comprehensibility of nature. . . . And its [science's] vocation will be ended as soon as the reduction of natural phenomena to simple forces is complete and the proof given that this is the only reduction of which phenomena are capable.

Helmholtz was expressing a pious hope, for even as these words were written, evidence was appearing that it is not possible to account for all phenomena in terms of masses responding to the action of simple and evident forces.

Today, if it was not obvious in the nineteenth century, we must face the failings of mechanism. Scientists are reasonably clear in the presentation of their creations, but they are clearest when they are clearly wrong. Until about the end of the nineteenth century they were sure that all the phenomena of nature would be explainable in mechanical terms. Those not already explainable soon would be. Notable among those still to be explained were the action of gravitation and of electromagnetic wave propagation.

As for gravitation, certainly scientists were aware that Newton had made strenuous efforts to explain the action of gravity. How did the sun's attraction act on planets millions and hundreds of millions of miles away? Newton's efforts failed, and he closed this endeavor with his famous "I frame no hypotheses." Mechanism did not aid him.

Why, then, did the eighteenth- and nineteenth-century scientists cling to mechanism? One answer is that hope springs eternal. More relevant is that they were so flushed with success in following Newton's lead that they lost sight of the problem of explaining the *physical* nature of gravitation. They resorted to the mathematical law of gravitation, and their successes (notably those of Lagrange and Laplace) in deducing some known irregularities in the heavenly motions and in encompassing new phenomena were so great, so remarkably accurate, that the problem of explaining the physical action of gravitation was buried under a heap of mathematical papers. We now know that gravitation is a scientific fiction suggested to an extent by the human ability to exert force.

Bishop Berkeley, in keeping with his general philosophy, attacked the concept of a physical force of gravitation. In his dialogue, *Alciphron* (1732), he wrote:

> EUPHRANOR: Let me entreat you Alciphron, be not amused by terms: lay aside the *word* force, and exclude every other thing from your thoughts, and then see what precise idea you have of force.
>
> ALCIPHRON: Force is that in bodies which produceth motion and other sensible effects.
>
> EUPHRANOR: It is then something distinct from those effects?
>
> ALCIPHRON: It is.
>
> EUPHRANOR: Be pleased now to exclude the consideration of its subject and effects, and contemplate force itself in its own precise idea.
>
> ALCIPHRON: I profess I find it no such easy matter.

> And that, replied Euphranor, which it seems that neither you nor I can frame an idea of, by your own remark of men's minds and faculties being made much alike, we may suppose others have no more idea than we.

In sum, the reliance on mathematical description even though physical understanding was completely lacking made possible not only Newton's amazing contributions but hundreds of succeeding ones. What these men did was to sacrifice physical intelligibility for mathematical description and mathematical prediction. In the words of English writer G. K. Chesterton (1874–1936), "We have seen the truth; and the truth makes no [physical] sense." As far as mechanism is concerned, the history of electromagnetic theory is roughly the same as that of gravitation. As we have noted, Faraday had introduced lines of force to explain various actions of electrical charges, magnetism, and the interaction of electrical charges. It was at least conceivable that the physical existence of these lines of force might someday be demonstrated. However, when Maxwell extended the action of electrical and magnetic phenomena to waves traveling hundreds and thousands of miles from one place to another, Faraday's lines of force proved to be totally inadequate even as a potential physical account. Instead, Maxwell accepted the ether, which had been advanced as the medium that carried light, as the medium that transmitted all electromagnetic waves, light included. Maxwell had made extensive efforts to supply a mechanical account of the transmission of electromagnetic waves, but these, like Newton's efforts to explain gravitation, ended in failure. Mathematical equations took over.

But in the light of more recent developments, mechanism or materialism is not tenable. Ether as a substance has been abandoned, and only mathematical laws "replace" it. Gravitational force has been replaced by relativistic geodesics in space–time. We accept the propagation of electromagnetic waves whose physical nature is unknown. We are also asked to accept a particle–wave duality that defies common sense, as if, by magic, electrons that are particles become waves when ejected from atoms. Relativity and quantum mechanics especially call for a profound revision of classical mechanics. These changes are not so disturbing if one traces the long history from Greek times onward to the rise of the classical mechanics of Newton, Lagrange, and Laplace. The revisions of Aristotelian and Scholastic mechanics and of Ptolemic astronomy were at least as revolutionary.

The inroads made on the mechanistic conception of nature by newer developments are evident in the plaintive remarks of Lord Kelvin (1824–1907), a leading figure in English scientific circles during the latter half of the nineteenth century:

> I am never content until I have constructed a mechanical model of the object I am studying. If I succeed in making one I understand; otherwise I do not. I wish to understand light as fully as possible, without introducing things that I understand still less.

However, Kelvin had to be content with far less "light" than he would have liked to possess.

Another doctrine that has been invoked many times throughout history to explain the behavior of nature is the notion of cause and effect. Thus we try to find causes, because this knowledge would enable us to bring about desired effects. Causality is a somewhat more vague doctrine than mechanism. It asserts cause and effect, although it does not insist on knowing the mechanism. For many centuries, roughly until 1900, causality was supported by the belief in mechanism. Many effects took place because between cause and effect a physical mechanism operated to produce the effect. Originally, causality implied contact between cause and effect, or spatial contiguity. However, this was soon extended to action at a distance, as in the case of gravitation.

Like most doctrines, causality had its origin in Greek thought. As discussed earlier, Aristotle distinguished four kinds of causes operating in the universe: formal causes that are plans or designs, final causes or purposes, material causes that reside in matter, and efficient causes that effect changes or happenings. Archimedes (287–212 B.C.), who was both a great mathematician and a scientist capable of applying his knowledge to practical affairs, emphasized a principle of causality akin to Aristotle's efficient cause, which brings about the result that matter behaves everywhere and at all times in an orderly and predictable manner.

The search for causes in modern science begins with Galileo. He did speak of the Earth's gravitational force as the cause of terrestrial motions, but he was compelled to ignore causality and had to be content with mathematical descriptions of the motions.

Newton and his contemporaries developed the concept, which was to remain essentially unchanged over the following two centuries, that causality is inherent in the nature of the physical world itself. It was in his search for causes that Newton introduced the universal force of gravitation as the cause of the elliptical motions of the planets, which might otherwise move along straight lines. Leibniz, too, did say that nothing happens without a cause; however, in his day, belief in cause and effect was just a belief.

A radically different understanding of cause and effect was advanced by Immanuel Kant. Strongly influenced by the rise of Newtonian science at a time when the cosmological theories of Descartes were still prevalent in Europe, he advocated Newton's system of celestial mechanics and even supplemented it substantially in his scientific treatise *A Theory of the Heavens* (1755). In his major philosophical work, *A Critique of Pure Reason* (1781), he asserted that causality was a logically necessary precondition for all rational thought. Therefore, it was in no need of support by factual evidence. In the second edition of the *Critique* (1787) he defined causality thus: "All changes happen according to a law of connection between cause and effect."

All of these conceptions of causality involve in various ways the idea of a nexus whereby the cause brings the effect into being. The Scottish philosopher David Hume (1711–1776) sought to purge causality of any metaphysical basis. In effect he questioned causality. In his important treatise on epistemology, *Enquiry Concerning the Human Understanding* (1793), he writes:

> The only immediate utility of all science is to teach us how to control and regulate future events by their causes. . . . Similar events are always conjoined with similar, of this we have experience; therefore we may define a cause to be an object followed by another and where all the objects similar to the first are followed by objects similar to the second.

In this wording, in which "object" might be better rendered as "event," Hume is saying that a situation C and a subsequent situation E are related as *cause* and *effect*, if the occurrence of C (or a situation similar to it) is always followed by E (or something similar), and if E never occurs unless C has occurred previously. Hume included the word similar in his definition because he wanted to make causality *experimentally verifiable;* and he realized, correctly, that a given situation cannot be found to reoccur if it is defined too precisely.

Having defined causality, he proceeded to attack it. It was his conviction that just because we have become aware of a particular sequence of cause and effect, even a very large number of times, this is no proof that the cause will be followed by the effect on future occasions. He concludes that our belief in causality is no more than a *habit* that he rightly avers is not an adequate basis for belief.

John Stuart Mill, the most celebrated English philosopher of the nineteenth century, reinforced Hume's denial of causality and added some ideas of his own. Mill gave his conception of causality in his *System of Logic* (1843) in these words: "The law of causation, the recognition of which is the main pillar of science, is but the familiar truth that invariability of succession is found by observation to obtain between every fact in nature and some other fact which has preceded it." Thus, like Hume, Mill makes "invariability of succession" the essence of causality, and again like Hume, he gives it an empirical basis. He strips causality of logical necessity and removes the idea of compulsion. He analyzes the circumstances under which he believes that one may assume the existence of a cause-and-effect relationship between two events: cause occurs spatially close to the event; cause is followed immediately by the event; and cause is always followed by the event. He does not explicitly refute Hume's contention that causality is a habit of thought. Instead, for Mill causality is an empirical generalization. Induction is the basis of some generalizations and in particular of the laws of nature. He does discuss the methods whereby a causal relationship may be inferred, for example, by the method of differences:

> If an instance in which the phenomenon under investigation occurs and an instance in which it does not occur have every circumstance in common save one, that occurring only in the former; the circumstance in which alone the two instances differ is . . . the cause, or an indispensable part of the cause, of the phenomenon.

This clearly stated principle is still used in many areas of science. Thus, an experiment performed on laboratory animals to test the effect of a new drug always involves two groups, chosen to be as nearly alike as possible in size, age, housing, feeding, and so on, with the single difference that one group receives the drug while the other, the *control* group, does not. In accordance with the method of differences, any effect observed in the former and not in the latter may be fairly taken to be caused by the drug.

A more devastating attack on causality was made by Bertrand Russell, English mathematician and philosopher and recipient of the 1950 Nobel Prize for literature. In a paper, "On the Notion of Cause," he says:

> All philosophers, of every school, imagine that causation is one of the fundamental axioms of science, yet, oddly enough, in advanced science, such as gravitational astronomy, the word "cause" never occurs. . . . The Law of Causality, I believe, like much that passes among philosophers, is a relic of a bygone age, surviving like the monarchy, only because it is erroneously supposed to do no harm.

Russell goes too far in his statement that causality is for science "a relic of a bygone age."

More recently, the theory of relativity upset the relationship of cause and effect. Under this relationship one usually assumes that the cause must precede the effect. According to relativity, however, the order of two events is no longer an absolute affair. When we discussed the question of simultaneity in Chapter IX, we found that the order of the two flashes of light depended on the observer. If these two flashes were replaced by events that appeared to be cause and effect to some observers, there might nevertheless be other observers who could not view the events in that relation, for to them the event called the effect might occur *before* the cause. Thus the conception of a cause-and-effect relationship is wanting.

In spite of its shortcomings, the principle of causality in science remained essentially unchanged throughout the classical period. And in spite of criticism from Hume, Mill, and Russell, toward the end of the nineteenth century causality was raised to the level of a self-evident truth, even as Kant had done a century before on metaphysical grounds. Typical of this attitude is the statement of Ludwig Boltzmann (1844–1906), given in his *Physiologische Optik:*

> The causal law bears the character of a purely logical law even in that the consequences derived from it do not really concern experience itself but the understanding thereof, and therefore it could never be refuted by any possible experience.

How the principle of causality fared with the development of quantum theory is a story we shall return to soon.

However, because the cause of an effect could not always be ascertained, as for example in the case of comets, and a mechanism could not always be found to explain various phenomena, a superseding doctrine did take hold in the nineteenth century—determinism. The distinction between causality and determinism had already been made by Descartes. The effect appears to follow the cause in time as a result of the limitations of human sense perceptions. *Causa sive ratio* (Cause is nothing but reason.) The meaning of this doctrine is best explained by an analogy. Given the axioms of Euclidean geometry, the properties of a circle (such as the circumference and the area) and the properties of inscribed angles are all immediately determined as necessary logical consequences. In fact, Newton is supposed to have asked why anyone bothered to write out the theorems of Euclidean geometry since they are obviously implied by the axioms. Most human beings, however, take a long time to discover each of these properties. But this discovery in time, which seems to relate axioms and theorems in the same temporal sequence as cause and effect, is illusory.

So it is with physical phenomena. To the divine understanding, all phenomena are coexisting and are comprehended in one mathematical structure. The senses, however, recognize events one by one and regard some as the causes of others. We can understand now, said Descartes, why mathematical prediction of the future is possible; it is because the mathematical relationships are preexisting. The mathematical relationship is the clearest physical explanation of a relationship. In brief, the real world is the totality of mathematically expressible motions of objects in space and time, and the entire universe is a great, harmonious, and mathematically designed machine. Moreover, many philosophers, including Descartes, insisted that these mathematical laws are fixed because God had so designed the universe and God's will is invariable. Whether or not humans could decipher God's will or penetrate God's design, the world functioned according to law, and lawfulness was undeniable, at least until the 1800s.

The Newtonian concept of a universe consisting of hard, indestructible particles acting on one another by well-determined, calculable forces was made the basis of a thorough and rigid determinism by the French astronomer and mathematician, the Marquis Pierre-Simon de Laplace (1749–1827). In his classic statement on the nature of determinism, he wrote:

An intelligence knowing, at any given instant of time, all forces acting in nature, as well as the momentary positions of all things of which the universe consists, would be able to comprehend the motions of the largest bodies of the world and those of the smallest atoms in one single formula, provided it were sufficiently powerful to subject all data to analysis; to it nothing would be uncertain, both future and past would be present before its eyes.

Actually Laplace's "one single formula" stretches the imagination. Determinists would be content with many formulas.

Laplace did not realize it, but he was writing the epitaph of mechanism and determinism. His concept involves a fanciful superhuman "intelligence," but the existence of such an intelligence is beside the point. If the universe does actually run inexorably in a rigidly deterministic manner through all of past and future time, it does so whether or not an intelligence knows about it, for in the universe of Laplace this knowledge exerts no influence. Because of his great and well-merited renown in the fields of mathematics and astronomy, Laplace's conception of a completely deterministic universe was widely discussed and was given great importance.

The deterministic point of view was held so firmly that philosophers applied it to the actions of human beings as part of nature. Ideas, volitions, and actions are necessary effects of matter acting on matter. The human will is determined by external physical and physiological causes. Hobbes explained apparent free will in this way: Events from without act on our sense organs, and these press on our brains. Motions within the brain produce what we call appetites, delights, or fears, but these feelings are no more than the presence of such motions. When appetite and aversion jostle each other, there is a physical state called deliberation. One motion prevails, and we say that we have exercised free will. However, no choice is really made by the individual. We are conscious of the result, but unconscious of the process that determined it. There is no free will. It is a meaningless conjunction of words. The will is bound fast in the actions of matter.

Voltaire in his *Ignorant Philosopher* states: "It would be very singular that all nature, all the planets, should obey eternal laws, and that there should be a little animal, five feet high, who, in contempt of these laws, could act as he pleased, solely according to his caprice." Chance, also, is nothing but a word invented to express the known effect of an unknown cause.

So disturbing was this conclusion that even materialists sought to modify its severity. Some said that, while human actions were determined, thoughts were not. The introduction of this dichotomy was not too comforting, for it meant that thoughts were useless in determining action and humans remained automatons. Others reinterpreted the meaning of freedom in an effort to retain some semblance of it. Voltaire hedged: "To be free means to be able to do what we like, not to be able to will what we like." Apparently, we must like what is willed for us in order to be free.

For scientific purposes, to say that event A determines event B means simply that, given A, it is possible to *calculate B*, and vice versa. Thus, with regard to the way determinism is *used* in an exact science, it may be characterized by the statement: Given the condition of a collection of objects at one particular instant, it is possible to determine by calculation its condition at any other instant, past or future.

The scientific conception of determinism is expressed best by functional relations between variables, that is, by formulas such as we have encountered in preceding chapters. Obviously, a functional relation carries no implication of cause and effect.

In fact, a good part of the business of an exact science is to establish functional relations between variables. When a relation of this kind has been found to be broadly valid and to express an important fact about the workings of the physical universe, it acquires the stature of a law of nature. Insofar as determinism is involved, the principle may be said to reside simply in the constancy and reliability of scientific laws. With due regard for the two facts that (1) the experimental data on which laws are based are never ideally precise, and (2) that all theoretical relations are tentative and subject to revision by new discoveries, determinism connotes no more and no less than the uniformity of nature.

However, determinism was not fated to endure. There are unstable factors—Clerk Maxwell called them singular points—in the operations of nature. A rock resting on the top of a mountain peak is unstable because a little push might start an avalanche. Similarly, the match that starts a forest fire, the little word that sets the world fighting, and the little gene that makes us philosophers or idiots are unstable phenomena. Such unstable factors are flaws in the deterministic world. Laws break down in these instances, and effects that are negligible under other circumstances can be dominating.

Maxwell cautioned his fellow scientists about the implications of these singular points:

> If, therefore, those cultivators of physical science . . . are led in the pursuit of the arcana of science to study the singularities and instabilities, rather than the continuities and stabilities of things, the promotion of natural

knowledge may tend to remove that prejudice in favor of determinism which seems to arise from assuming that the physical science of the future is a mere magnified image of that of the past.

The leader of his own generation was actually the prophet of the next one. Some of Maxwell's own contributions to the theory of gases helped to prepare the way for the demise of determinism. The cracks or flaws he saw in this scheme of things soon widened, and the deterministic world fell apart.

Determinism had to yield to statistical laws. Before we pursue this concept, let us see what we mean by such laws. An example of problems that have been successfully attacked by the theories of statistics and probability will demonstrate the effectiveness of this mathematical approach. The largest business in the United States is insurance. It is obvious that any attempt to predict the year of death of any one individual by deduction from first principles would be doomed to failure. Yet by obtaining data on the life spans of thousands of individuals and by applying the theory of probability, the insurance companies are able to insure people at rates that are fair to the individual who pays the premiums and to the companies that take the risk.

The use of statistical laws in physics started with statistical mechanics, where one could at least believe that if we could embrace millions of collisions of molecules behaving deterministically, we could determine, say, the behavior of a gas; however, the number is so large that it is impossible to consider the mass effect except by statistical means. The first major use of statistical laws was made by Ludwig Boltzmann in his work on gases. This was a radical step in a world seemingly at ease with mechanism and determinism, and it caused fierce debates. However, Boltzmann insisted that the task of physics is not to summon empirical data to the judgment of our laws and thoughts but to adapt our thoughts, ideas, and concepts to what is empirically given. Boltzmann's statistical mechanics was in his time derided as the speculation of a "mathematical terrorist."

Radioactivity, the seemingly arbitrary behavior of electrons as waves and particles, and the rather unpredictable ejection of particles from the nuclei of atoms certainly challenged determinism. Moreover, the behavior of Planck's quanta, Einstein's photons, and Bohr's jumps of electrons cannot be predicted with certainty.

The indeterminacy principle proclaimed by Werner K. Heisenberg (1901–1976) in 1927 (see Chapter X) also played an important role in shaking the belief in determinism. In an article published in 1927 Heisenberg attacked both causality and determinism:

But in the strong formulation of the causal law, if we know the present exactly then we can calculate the future; it is not the final cause which is wrong, but the assumption. It is impossible for us in principle to know the present in all its determined pieces. Therefore, all perception is a selection from among a large number of possibilities and a restriction on future possibilities. As the statistical character of the quantum theory is so closely linked to the imprecision of all perception, one is tempted to suspect that another "real" world is hidden behind the perceived, statistical world in which the causal law is valid. But such speculations appear to us . . . pointless and sterile. Physics must give only a formal description of the connection between perceptions. A much better description of the real facts is: because all experiments are subject to the laws of quantum mechanics, quantum mechanics definitely shows the invalidity of the causal law.

Heisenberg's indeterminacy principle does not merely state that the causal links of quantum phenomena are beyond our powers of detection; it clearly implies that these links do not exist. This was Heisenberg's own inference. In view of the indeterminacy principle, causality and determinism become meaningless. Quantum mechanics can now only be a statistical discipline. It presents no exact description of an individual particle and makes no exact prediction of its behavior. However, it can make very accurate predictions regarding the behavior of large aggregates of particles.

Richard von Mises and others who wrote about quantum theory advanced indeterminate mechanism. All determinate laws became viewed as nothing more than approximate and purely passive reflections of the probabilistic relationships associated with the laws of chance. Individual processes and events in the atomic domain are thus completely lawless. As Eddington predicted in his *The Nature of the Physical World* (1933), "science has made determinism untenable."

In 1957 Hans Reichenbach in his *Atom and Cosmos* stressed that the probabilistic interpretation of all physical results is the correct one. What is most probable is what happens within the limits of observation. Only on the large scale, where innumerable atoms combine in processes of great probability, can we treat such phenomena as certain in practice. Basically, however, even large-scale events are probabilistic. The concepts of space, time, substance, force, causality, and laws are borrowed from ordinary human experience in the "middle-sized" world but are certainly not appropriate for atomic phenomena.

For a long time other important physicists such as Born, Bohr, and Pauli held the view, with minor variations, that the happenings of nature were subject only to the probabilistic interpretation, while

Planck, Einstein, von Laue, de Broglie, Schrödinger, and others did not agree; they adhered to the classical mechanical conceptions of causality and determinism. The essence of the dispute was primarily whether the statistical character of the laws of quantum physics was a temporary expedient due to our lack of knowledge, which would be superseded in the course of time by laws like those of Newtonian mechanics, or whether the statistical laws had an objective character— that is, independent of our knowledge and consciousness—and corresponded to the actual happenings in the microworld.

Most of us are familiar with Einstein's view that God does not throw dice. He expressed this conviction in two letters that are reproduced by Ronald W. Clark in *Einstein: The Life and Times*. The first letter, addressed to Max Born in 1926, says:

> Quantum mechanics is certainly imposing. But an inner voice tells me that it is not yet the real thing. The theory says a lot, but does not bring us any closer to the secret of the Old One. I, at any rate, am convinced that He does not throw dice.

The second letter, written much later to James Franck, states:

> I can, if the worst comes to the worst, still realize that the Good Lord may have created a world in which there are no natural laws. In short, a chaos. But that there should be statistical laws with definite solutions, e.g., laws which compel the Good Lord to throw dice in each individual case, I find highly disagreeable.

And in another passage, from *The World as I See It* (1934), Einstein said, "God is subtle; He is not malicious." Furthermore, Einstein and his colleagues added, in a joint paper in the *Physical Review* of 1935, that the theory of wave mechanics is incomplete. Einstein said statistical quantum theory would in the future be like statistical mechanics wherein the motion of individual particles (for example, molecules of a gas) are determined, but because there are so many, statistics and probability are applied. And Paul A. M. Dirac, the British physicist who contributed much to the new physics, takes the same position (1978):

> I think it might turn out that ultimately Einstein will prove to be right, because the present form of quantum mechanics should not be considered as the final form. . . . I think that it is quite likely that at some future time we may get an improved quantum mechanics in which there will be a return to determinism and which will therefore justify the Einstein point of view. But such a return to determinism could only be made at the

expense of giving up some other basic idea which we now assume without question. We would have to pay for it in some way which we cannot yet guess, if we are to re-introduce determinism.

Dirac certainly seems correct in pointing out that some ideological barrier is diverting us from the development of a more complete deterministic theory. As Alexander Pope put it in his *Essay on Man,* "All chance, direction which thou cans't not see. . . . "

Neither Dirac nor Einstein proposed an alternative model to meet this need. Other physicists, such as David Bohm (1957) and Shoichi Sakata (1978), who criticized probabilistic quantum mechanics, have also failed to propose useful alternative models. Many other able scientists have tackled the same problem without results. However, quantum mechanics is so fully developed by now that the answer can hardly depend on more experimental data.

Although scientists still use the deterministic laws of classical mechanics to deal with events about objects that can be readily seen or handled, the middle-sized phenomena that Reichenbach spoke of, their attitude toward determinism in such events has been profoundly altered because of the new insights afforded by quantum mechanics. Things are thought to happen the way they do because it is highly probable that they do so and highly improbable that they do otherwise.

Mechanism, causality, and determinism are three of the many philosophies of science that have been profoundly affected by recent scientific creations. There are many more. Let us look briefly at one or two others.

The philosophy of idealism is another way of solving the metaphysical problem of our relationship to an external world. Idealism solves the problem by lopping off one end of it—by denying as Berkeley did the existence of an external world (see "Historical Overview"). All our awareness of the external world actually occurs within ourselves; hence, the belief that this awareness is generated by objects external to ourselves may well be an illusion. If we look at a tree, it surely exists in our consciousness. If we turn away, the tree no longer exists there. If we remember it, or if we hear some other person assuring us that it is still there, we are again experiencing nothing but mental processes.

The common intuitive reaction to idealism is to dismiss it as absurd. The redoubtable Dr. Samuel Johnson (1709–1784) thought to refute it by kicking a large rock with his foot. In spite of many attempts by competent philosophers, idealism has never been finally refuted.

Because the existence of something that is not causing sensory perceptions in any conscious being is impossible to prove experimentally, physical existence independent of humanity ought to be declared meaningless; furthermore, all scientists should be idealists. Nevertheless, all of classical science has been based solidly on the premise that an external objective universe does exist. Scientists are generally agreed that nature is not deceiving them and that their conception of a real external world is justified.

The classical scientist, if challenged on his or her belief in an objective universe, would reply that one's observations do not have any appreciable effect on the observed object. This scientist would assert that one is actually determining what the object was like before and what it also will be like after observing it. However, this presupposition of classical science is no longer tenable. Observations do have an effect on the objects observed, and this effect is by no means unappreciable for the elementary constituents of the universe. Heisenberg made this point.

Classical science had assumed a priori that an external world exists. The mathematical equations of classical mechanics were thought to describe what is actually happening in this external world. Quantum mechanics also has its mathematical equations, but these are a description of observations, not of the actual particles but of the effect of these particles on screens somewhat like television screens.

In opposition to idealism, the philosophy of logical positivism asserts that truths are built up only on observable facts. The positivists are anti-metaphysical, and their only source of meaningful knowledge is experience. From experience one obtains the basic assertions that can then be elaborated by rigorous deduction. However, the meaning of any proposition is identical with the means of verifying it. John Stuart Mill is a representative of the positivist philosophy, and he, too, asserts that, while knowledge comes primarily through the senses, it also includes the relations the conscious mind formulates concerning the evidence afforded by the senses, such as scientific laws. Furthermore, while the positivists agree with the idealists that there is no way of proving that an external world exists, they insist that there is also no proof that it does not. The positivists are basically empiricists who draw a sharp distinction between experiences and objects of reason, and deny the reality of the latter.

What have we learned in this brief account? Our objective has been a simple one—to indicate how recent developments in physical science cause us constantly to reexamine firmly held views, to note the changes these developments make in our own lives and in the way we view nature. Philosophies of science or, if one prefers, of nature's behavior are broad generalizations based on current scientific knowledge. As our knowledge changes from one epoch to another, the philosophies must be altered; therefore, we should never lose sight of the "hard core" of *scientific* findings.

The objective of this book has been to make evident how much these findings have been largely and in some areas completely dependent on mathematics. Because mathematics, despite some contrary views, is surely a human creation, what can we conclude? Whether nature is orderly, designed, and even purposeful (as Aristotle would have it) cannot be decided. What is certain is that humanity's most powerful tool, mathematics, is providing some understanding and some mastery of the complex and bewildering natural phenomena.

Bibliography

This Bibliography suggests references for additional reading on several levels. The reason is simply that readers of the text may be well informed on some topics, and wish to pursue these still further, or that others would feel the need for more elementary presentations.

GENERAL

Bradley, F.H. *Appearance and Reality,* 2nd ed. New York: Oxford University Press, 1969.

Broad, C.D. *Scientific Thought.* Paterson, N.J.: Littlefield, Adams & Co., 1959.

Bronowski, J. *The Common Sense of Science.* London: William Heineman, 1951.

D'Abro, A. *The Decline of Mechanism in Modern Physics.* New York: D. Van Nostrand Co., 1939.

Di Francia, G.T. *The Investigation of the Physical World.* New York: Cambridge University Press, 1981.

Eddington, A.S.: *The Nature of the Physical World.* New York: Macmillan, 1933.

Einstein, A. *Essays in Science.* New York: Philosophical Library n.d.

Hobson, E.W. *The Domain of Natural Science.* New York: Dover Publications, 1968.

Jeans, Sir J. *The Growth of Physical Science.* New York: Cambridge University Press, 1951.

———. *The Mysterious Universe.* New York: Macmillan, 1930.

———. *The New Background of Science.* New York: Macmillan, 1933.

———. *Physics and Philosophy.* Ann Arbor: The University of Michigan Press, 1958.

Johnson, M. *Science and the Meaning of Truth.* London: Faber & Faber, 1946.

Kemble, E.C. *Physical Science, Its Structure and Development.* Cambridge, Mass.: The M.I.T. Press, 1966.

Kline, M. *Mathematical Thought from Ancient to Modern Times.* New York: Oxford University Press, 1972.

———. *Mathematics and the Physical World.* New York: Dover Publications, 1981.

Körner, S. *Experience and Theory.* New York: The Humanities Press, 1966.

Koyré, A. *From the Closed World to the Infinite Universe.* Baltimore: The Johns Hopkins Press, 1957.

Lakatos, I. *Mathematics, Science and Epistemology,* 2 vols. New York: Cambridge University Press, N.Y., 1978.

Margenau, H. *The Nature of Physical Reality.* New York: McGraw-Hill, 1950.

Marion, J. B. *Physics in the Modern World.* New York: Academic Press, 1976.

Munn, A. M. *From Nought to Relativity.* London: George Allan & Unwin, 1973.

Peierls, R.E. *The Laws of Nature.* New York: Charles Scribner's Sons, 1956.

Reichenbach, H. *Experience and Prediction.* Chicago: The University of Chicago Press, 1938.

Schrödinger, E. *Mind and Matter.* New York: Cambridge University Press, 1958.

Sutton, O.G.: *Mathematics in Action.* London: G. Bell & Sons, 1954.

Weyl, H. *Philosophy of Mathematics and Natural Science.* Princeton, N.J.: Princeton University Press, 1949.

Whitehead, A.N.: *Science and the Modern World.* London: Cambridge University Press, 1953.

Whittaker, Sir E. *From Euclid to Eddington.* New York: Dover Publications, 1958.

HISTORICAL OVERVIEW: IS THERE AN EXTERNAL WORLD?

Baum, R.J. *Philosophy and Mathematics.* San Francisco: Freeman, Cooper & Co., 1973.

Benaceraff, P. and Putnam, H. *Selected Readings.* Englewood Cliffs, N.J.: Prentice-Hall, 1964.

Berkeley, G. *Three Dialogues between Hylas and Philonous.* Chicago: The Open Court Publishing Co., 1929.

Cassirer, E. *Substance and Function.* New York: Dover Publications, 1953.

Kant, I. *Critique of Pure Reason,* many editions available.

Körner, S. *The Philosophy of Mathematics.* London: Hutchinson University Library, 1960.

Lindsay, R.B. *The Nature of Physics.* Providence, R.I.: Brown Unviersity Press, 1968.

Reichenbach, H. *Experience and Prediction.* Chicago: The University of Chicago Press, 1938.

Russell, B. *A History of Western Philosophy.* New York: Simon & Schuster, 1945.

————. *Our Knowledge of the External World.* New York: The New American Library, 1956.

Urmson, J.O. *Berkeley.* New York: Oxford University Press, 1983.

Warnock, G.J. *Berkeley.* Notre Dame, Ind.: University of Notre Dame Press, 1983.

Whitehead, A.N. *Science and the Modern World.* London: Cambridge University Press, 1953.

Wolgast, E.H. *Paradoxes of Knowledge.* Ithaca, N.J.: Cornell University Press, 1977.

CHAPTER I

Attneave, F. "Multistability in Perception." *Scientific American,* December 1971, pp. 42–71.

Battersby, M. *Trompe L'Oeil, the Eye Deceived.* New York: St. Martin's Press, 1974.

Carraher, R.G., and Thurston, J.B. *Optical Illusions and the Visual Arts.* New York: Van Nostrand Reinhold Co., 1966.

Gibson, J.J. *The Perception of the Visual World.* New York: Houghton-Mifflin Co., 1950.

Gillam, B. "Geometrical Illusions." *Scientific American,* January 1980, pp. 102–111.

Gilson, E. *Painting and Reality.* New York: Pantheon Books, 1975.

Gombrich, E.H. *Art and Illusion,* 2nd ed. New York: Pantheon Books, 1961.

Gregory, R.L. *The Intelligent Eye.* New York: McGraw-Hill, 1970.

————. "Visual Illusions." *Scientific American,* November 1968, pp. 66–76.

Helmholtz, H. von. *On the Sensations of Tone.* New York: Dover Publications, 1954.

Ittelson, W.H., and Kilpatrick, F.P. "Experiments in Perception." *Scientific American,* August 1951, pp. 50–55.

Luckiesh, M. *Visual Illusions.* New York: Dover Publications, 1965.

Maurois, A. *Illusions.* New York: Columbia University Press, 1968.

Murch, G.M., ed. *Studies in Perception.* New York: The Bobbs Merrill Co., 1976.

Rock, Irvin. *Perception.* Holmes, Pa.: Scientific American Library, 1983.

Tolansky, S. *Curiosities of Light Rays and Light Waves.* New York: Elsevier Publishing Co., 1965.

————. *Optical Illusions.* New York: Pergamon Press, 1964.

CHAPTER II

Black, M. *The Nature of Mathematics.* New York: Harcourt, Brace, Jovanovich, 1935.

Bourbaki, N. "The Architecture of Mathematics." *American Mathematical Monthly* 57 (1950):221–232.

Courant, R. "Mathematics in the Modern World." *Scientific American,* September 1964, pp. 40–49.

Dyson, F.J. "Mathematics in the Physical Sciences." *Scientific American,* September 1964, pp. 129–146.

Eves, H., and Newsom, C.V. *An Introduction to the Foundations and Fundamental Concepts of Mathematics,* rev. ed. New York: Holt, Rinehart & Winston, 1965.

Goodman, N.D. "Mathematics as an Objective Science." *American Mathematical Monthly* 86 (1979):540–551.

Goodstein, R.L. *Essays in the Philosophy of Mathematics.* Leicester, England: Leicester University Press, 1965.

Hamilton, A.G. *Numbers, Sets and Axioms.* New York: Cambridge University Press, 1983.

Körner, S. *The Philosophy of Mathematics.* London: Hutchinson University Library, 1960.

Polya, G. *Mathematical Methods in Science.* Washington, D.C.: The Mathematical Association of America, 1977.

Steiner, M. *Mathematical Knowledge.* Ithaca, N.Y.: Cornell University Press, 1975.

Titchmarsh, E.C. *Mathematics for the General Reader.* New York: Dover Publications, 1981.

Walker, M. *The Nature of Scientific Thought.* Englewood Cliffs, N.J.: Prentice-Hall, 1963.

Whitehead, A.N. *An Introduction to Mathematics.* New York: Henry Holt & Co., 1939.

Whitney, H. "The Mathematics of Physical Quantities." *American Mathematical Monthly* 75 (1968):115–138 and 227–256.

Wilder, R.I. *Introduction to the Foundations of Mathematics.* New York: John Wiley & Sons, 1965.

CHAPTER III

Apostle, H.G. *Aristotle's Philosophy of Mathematics,* Chicago: The University of Chicago Press, 1952.

Clagett, M. *Greek Science in Antiquity.* New York: Abelard-Schumann, 1955.

Sambursky, S. *The Physical World of the Greeks.* London: Routledge & Kegan Paul, 1956.

Schrödinger, E. *Nature and the Greeks.* New York: Cambridge University Press, 1954.

CHAPTER IV

Armitage, A. *Copernicus, the Founder of Modern Astronomy.* London: George Allen & Unwin, 1938.

————. *John Kepler.* London: Faber & Faber, 1966.

————. *The World of Copernicus.* New York: The New American Library, 1951.

Baumgardt, C. *Johannes Kepler: Life and Letters.* London: Victor Gollancz, 1952.

Berry, A. *A Short History of Astronomy.* New York: Dover Publications, 1961.

Boas, M. *The Scientific Renaissance 1450–1630.* London: Collins, 1962.

Caspar, N. *Kepler.* New York: Collier Books, 1962.

De Santillana, G. *The Crime of Galileo.* Chicago: The University of Chicago Press, 1955.

Dreyer, J.L.E. *A History of Astronomy from Thales to Kepler.* New York: Dover Publications, 1953.

Galilei, G. *Dialogue on the Great World Systems.* Chicago: The University of Chicago Press, 1953.

Gingerich, W. "The Galileo Affair." *Scientific American,* August 1983, pp. 132–143.

Kuhn, T.S. *The Copernican Revolution.* Cambridge, Mass.: Harvard University Press, 1957.

Margenau, H. *The Nature of Physical Reality.* New York: McGraw-Hill, 1950.

CHAPTER V

Burtt, E.A. *The Metaphysical Foundations of Modern Physical Science,* rev. ed. London: Routledge & Kegan Paul, 1932.

Butterfield, H. *The Origins of Modern Science.* New York: Macmillan 1932.

————. "The Scientific Revolution." *Scientific American,* September, 1960, pp. 173–192.

Clavelin, M. *The Natural Philosophy of Galileo.* Cambridge, Mass.: The M.I.T. Press, 1974.

Crombie, A.C. *Augustine to Galileo.* London: Falcon Press, 1952.

Dampier-Whetham, W.C.D. *A History of Science and Its Relations with Philosophy and Religion.* New York: Cambridge University Press, 1929.

Doney, W., ed. *Descartes, a Collection of Critical Essays.* Notre Dame, Ind.: The University of Notre Dame Press, 1968.

Drabkin, I.E., and Drake, S. *Galileo Galilei: On Motion and Mechanics.* Madison: The University of Wisconsin Press, 1960.

Drake, S. *Discoveries and Opinions of Galileo.* New York: Doubleday & Co., 1957.

Eaton, R.M. *Descartes Selections.* New York: Charles Scribner's Sons, 1927.

Galilei, G. *Dialogues Concerning Two New Sciences.* New York: Dover Publications, 1952.

————. *On Motion and on Mechanics.* Madison: The University of Wisconsin Press, 1960.

Hall, A. R. *From Galileo to Newton 1630–1720.* London: Collins, 1963.

————. *The Scientific Revolution.* New York: Longmans Green & Co., 1954.

Hooykass, R. *Religion and the Rise of Modern Science.* Edinburgh: Scottish Academic Press, 1972.

Randall, J.H., Jr. *The Making of the Modern Mind,* rev. ed. New York: Houghton-Mifflin Co., 1940.

Redwood, J. *European Science in the Seventeenth Century.* New York: Barnes & Noble, 1977.

Rée, J. *Descartes.* London: Allan Lane, 1974.

Scott, J.F. *The Scientific Work of René Descartes.* London: Taylor & Francis, 1952.

Strong, E.W. *Procedures and Metaphysics.* Berkeley: The University of California Press, 1936.

Vrooman, J.R. *René Descartes, a Biography.* New York: G.P. Putnam's Sons, 1970.

Wolf, A. *A History of Science, Technology and Philosophy in the 16th and 17th Centuries,* 2nd ed. London: George Allen & Unwin, 1950.

CHAPTER VI

Andrade, E.N. da C. *Sir Isaac Newton, His Life and Work.* New York: Doubleday & Co., 1954.

Bell, A.E. *Newtonian Science.* London: Edward Arnold, 1961.

Calder, N. *The Comet Is Coming!* New York: The Viking Press, 1981.

Cohen, I.B. *Introduction to Newton's Principia.* Cambridge, Mass.: Harvard University Press, 1971.

————. *Sir Isaac Newton's Papers and Letters on Natural Philosophy.* Cambridge, Mass.: Harvard University Press, 1958.

De Morgan, A. *Essays on the Life and Work of Newton.* Chicago: The Open Court Publishing Co., 1914.

Dijksterhuis, E.J. *The Mechanization of the World Picture.* New York: Oxford University Press, 1961.

Grosser, M. *The Discovery of Neptune.* New York: Dover Publications, 1979.

Hall, A.R. *From Galileo to Newton 1630-1720.* London: Collins, 1963.

Hesse, M.B. *Forces and Fields.* New York: Philosophical Library, 1962.

Jammer, M. *Concepts of Force.* Cambridge, Mass.: Harvard University Press, 1957.

Kline, M. *Mathematics, the Loss of Certainty.* New York: Oxford University Press, 1980.

More, L.T. *Isaac Newton, a Biography.* New York: Dover Publications, 1962.

Newton, Sir I. *Mathematical Principles of Natural Philosophy,* 3rd ed. Berkeley: The University of California Press, 1946.

Palter, R., ed.: *The Annus Mirabilis of Sir Isaac Newton, 1666-1966.* Cambridge, Mass.: The M.I.T. Press, 1970.

Slichter, C.S. "The Principia and the Modern Age." *American Mathematical Monthly,* August–September 1937, pp. 433–444.

Thayer, H.S., ed. *Newton's Philosophy of Nature.* New York: Hafner Publishing Co., 1953.

Valens, E.G. *The Attractive Universe.* Cleveland, Oh.: The World Publishing Co., 1969.

Westfall, R.S. *Never at Rest, a Biography of Sir Isaac Newton.* New York: Cambridge University Press, 1980.

CHAPTER VII

Bromberg, J.L. "Maxwell's Displacement Current and His Theory of Light." *Archive for History of Exact Sciences* 4 (1967):218–234.

Campell, L., and Garnett, W. *The Life of James Clerk Maxwell.* New York: Johnson reprint, 1969.

Domb, G. ed. *Clerk Maxwell and Modern Science.* London: The Athlone Press, 1963.

Everitt, C.W.F. *James Clerk Maxwell, Physicist and Natural Philosopher.* New York: Charles Scribner's Sons, 1975.

Haas-Lorentz, G.L., ed.: *H.A. Lorentz: Impressions of His Life and Work.* Amsterdam: North-Holland Publishing Co., 1957.

MacDonald, D.K.C. *Faraday, Maxwell, and Kelvin.* New York: Doubleday & Co., 1964.

Newton, Sir. I. *Opticks.* New York: Dover Publications, 1952.

Skilling, H.H. *Fundamentals of Electric Waves.* New York: John Wiley & Sons, 1942.

Thomson, Sir J.J., et al. *James Clerk Maxwell, A Commemoration Volume, 1831-1931.* New York: Cambridge University Press, 1931.

Whittaker, Sir. E. *A History of the Theories of Aether and Electricity,* 2 vols. London: Thomas Nelson & Sons, 1951 and 1953.

CHAPTER VIII

Bonola, R. *Non-Euclidean Geometry.* New York: Dover Publications, 1955.

Faber, R.L. *Differential Geometry and Relativity Theory.* New York: Marcel Dekker, 1983.

Golos, E.B. *Foundations of Euclidean and Non-Euclidean Geometry.* New York: Holt, Rinehart & Winston, 1968.

Greenberg, M.J. *Euclidean and Non-Euclidean Geometries.* San Francisco: W.H. Freeman & Co., 1974.

Kline, M. *Mathematics, the Loss of Certainty.* New York: Oxford University Press, 1980.

Wolfe, H.E. *Introduction to Non-Euclidean Geometry.* New York: The Dryden Press, 1945.

CHAPTER IX

Bergmann, P.G. *The Riddle of Gravitation.* New York: Charles Scribner's Sons, 1968.

Bondi, H. *Relativity and Common Sense.* New York: Dover Publications, 1980.

Born, M. *Einstein's Theory of Relativity.* New York: Dover Publications, 1962.

Calder, N. *Einstein's Universe.* New York: Greenwich House, 1982.

Coleman, J.A. *Relativity for the Layman.* New York: The New American Library, 1954.

D'Abro, A. *The Evolution of Scientific Thought.* New York: Dover Publications, 1949.

Davies, P.C.W. *Space and Time in the Modern Universe.* New York: Cambridge University Press, 1977.

Clarke, C. *Elementary General Relativity.* New York: John Wiley & Sons, 1980.

Eddington, A.S. *The Mathematical Theory of Relativity.* New York: Cambridge University Press, 1960.

————. *Space, Time and Gravitation.* New York: Cambridge University Press, 1953.

Einstein, A. *The Meaning of Relativity.* Princeton, N.J.: Princeton University Press, 1945.

————. *Relativity the Special and the General Theory.* New York: Crown Publishers, 1961.

————. *Sidelights on Relativity.* New York: Dover Publications, 1983.

————, and Infeld, L. *The Evolution of Physics.* New York: Simon & Schuster, 1938.

Faber, R.L. *Differential Geometry and Relativity Theory.* New York: Marcel Dekker, 1983.

Frankel, T. *Gravitational Curvature.* San Francisco: W.H. Freeman & Co., 1979.

Machamer, P.K., and Turnbull, R.G., eds. *Motion and Time, Space and Matter.* Columbus: Ohio State University Press, 1976.

Nevanlinna, R. *Space Time and Relativity.* Reading, Mass.: Addison-Wesley Publishing Co., 1968.

Pais, A. *"Subtle Is the Lord," The Science and the Life of Albert Einstein.* New York: Oxford University Press, 1982.

Pyenson, L. "Hermann Minkowski and Einstein's Special Theory of Relativity." *Archive for History of Exact Sciences* 17 (1977):71–95.

Reichenbach, H. *From Copernicus to Einstein.* New York: Dover Publications, 1980.

————. *The Philosophy of Space and Time.* New York: Dover Publications, 1957.

Rindler, W. *Essential Relativity, Special, General and Cosmological.* New York: Van Nostrand, 1969.

Rucker, R.V.B. *Geometry, Relativity and the Fourth Dimension.* New York: Dover Publications, 1977.

Russell, B. *The ABC of Relativity.* New York: Harper & Brothers, 1926.

Schild, A. "The Clock Paradox in Relativity Theory." *American Mathematical Monthly* 66 (1959):1–18.

Schilpp, P.A., ed. *Albert Einstein: Philosopher Scientist.* New York: Harper & Row, 1959.

Shankland, R.S. "The Michelson–Morley Experiment." *Scientific American,* November 1964, pp. 107–114.

CHAPTER X

Andrade, E.N. da C. "The Birth of the Nuclear Atom." *Scientific American,* November 1956, pp. 93–104.

Audi, M. *The Interpretation of Quantum Mechanics.* Chicago: The University of Chicago Press, 1973.

Baker, A. *Modern Physics and Antiphysics.* Reading, Mass.: Addison-Wesley Publishing Co., 1970.

Bertsch, G.F. "Vibrations of the Atomic Nucleus." *Scientific American,* May 1983, pp. 62–73.

Bloom, E.D., and Feldman, G.J. "Quarkonium." *Scientific American,* May 1982, pp. 66–77.

Born, M. *The Restless Universe.* New York: Harper & Brothers, 1936.

Burbridge, G., and Hoyle, F. "Anti-Matter." *Scientific American,* April 1958, pp. 34–39.

Cerny, J., and Poskanzer, A.M. "Exotic Light Nuclei." *Scientific American,* June 1978, pp. 60–72.

De Broglie, L. *Physics and Microphysics.* New York: Pantheon Books, 1955.

————. *The Revolution in Physics.* London: Routledge & Kegan Paul, 1954.

d'Espagnat, B. "The Quantum Theory and Reality." *Scientific American,* November 1979, pp. 158–181.

Feinberg, G. *What Is the World Made of? Atoms, Leptons, Quarks, and Other Tantalizing Particles.* New York: Doubleday Anchor Press, 1977.

Fritzsch, H. *Quarks the Stuff of Matter.* New York: Basic Books, 1983.

Gamow, G. "The Principle of Uncertainty." *Scientific American,* January 1958, pp. 51–57.

————. *Thirty Years That Shook Physics, the Story of Quantum Theory.* New York: Doubleday & Co., 1966.

Gell-Mann, M., and Rosenbaum, E.P. "Elementary Particles." *Scientific American,* July 1957, pp. 72–88.

Guillemin, V. *The Story of Quantum Mechanics.* New York: Charles Scribner's Sons, 1968.

Heisenberg, W.K. *The Physical Principles of the Quantum Theory.* New York: Dover Publications, 1949.

————. *Physics and Philosophy.* New York: Harper & Brothers, 1958.

Hoffman, B. *The Strange Story of the Quantum.* New York: Dover Publications, 1959.

Hund, F. *The History of Quantum Theory.* New York: Barnes & Noble, 1974.

Mistry, N.B., et al. "Particles with Naked Beauty." *Scientific American,* July 1983, pp. 106–115.

Mulvey, J.H., ed. *The Nature of Matter.* New York: Oxford University Press, 1981.

Pagels, H.R. *The Cosmic Code.* New York: Simon & Schuster, 1982.

Perlman, J.S. *The Atom and the Universe,* Wadsworth Publishing Co., Belmont, Calif., 1970.

Reichenbach, H. *Atom and Cosmos, the World of Modern Physics.* New York: George Braziller, 1957.

Rusk, R.D. *Introduction to Atomic and Nuclear Physics.* New York: Appleton-Century-Crofts, 1958.

Rydnick, V. *ABC's of Quantum Mechanics,* translation, Moscow: MIR Publishers, 1978.

Schrödinger, E. "What Is Matter?" *Scientific American,* September 1953, pp. 52–57.

Segrè, E., and Wiegand, C.E. "The Antiproton." *Scientific American,* June 1956, p. 37 ff.

Slater, J.C. *Concepts and Development of Quantum Physics.* New York: Dover Publications, 1969.

Smorodinsky, Ya.A. *Particles, Quanta, Waves.* Moscow: MIR Publishers, 1976.

Trefil, J.S. *From Atoms to Quarks.* New York: Charles Scribner's Sons, 1980.

Weinberg, S. "The Decay of the Proton." *Scientific American,* June 1981, pp. 64–75.
———. *The Discovery of Subatomic Particles.* New York: Scientific American Library, 1983.
Weisskopf, V.F. *Physics in the Twentieth Century.* Cambridge, Mass.: The MIT Press, 1972.
Whittaker, Sir E. *A History of the Theories of Aether and Electricity,* vol. 2. London: Thomas Nelson & Sons, 1953.
Woodgate, G.K. *Elementary Atomic Structure.* New York: Oxford University Press, 1983.
Zukav, G. *The Dancing Wu Li Masters.* New York: Wm. Morrow & Co., 1979.

CHAPTER XI

Birkhoff, G.D. "The Mathematical Nature of Physical Theories." *American Scientist* 31 (1943):281–310.
Bohr, N. *Atomic Physics and Human Knowledge.* New York: John Wiley & Sons, 1958.
Braithwaite, R.B. *Scientific Explanation.* New York: Cambridge University Press, 1953.
Bridgman, P.W. *The Logic of Modern Physics.* New York: Macmillan, 1946.
———. *The Nature of Physical Theory.* Princeton, N.J.: Princeton University Press, 1936.
Browder, F.E. "Does Pure Mathematics Have a Relation to the Sciences?" *American Scientist* 64 (1976):542–549.
Buchanan, S. *Truth in the Sciences.* Charlottesville, Va.: University Press of Virginia, 1972.
De Broglie, L. "The Role of Mathematics in the Development of Contemporary Theoretical Physics." In *Great Currents of Mathematical Thought,* vol. 2, edited by F. Le Lionnais, pp. 78–93. New York: Dover Publications, 1971.
Dyson, F.J. "Mathematics in the Physical Sciences." *Scientific American,* September 1964, pp. 129–146.
Goodstein, R.L. *Essays in the Philosophy of Mathematics.* Leicester, England: Leicester University Press, 1965.
Hesse, M.B. *Science and the Human Imagination.* London: SCM Press, 1954.
Jammer, M. *Concepts of Space.* Cambridge, Mass.: Harvard University Press, 1954.
Lanczos, C. *Space through the Ages.* New York: Academic Press, 1970.
Russell, B. *The Scientific Outlook.* New York:W.W. Norton and Co., 1962.
Stewart, I. "The Science of Significant Form." *Mathematical Intelligencer* 3 (1981):50–58.

CHAPTER XII

Barrett, W. *The Illusion of Technique.* New York: Doubleday & Co., 1979.
Bridgman, P.W. *The Logic of Modern Physics.* New York: Macmillan, 1946.
———. *The Nature of Physical Theory.* Princeton, N.J.: Princeton University Press, 1936.
Bunge, M. *The Myth of Simplicity.* Englewood Cliffs, N.J.: Prentice-Hall, 1963.
Einstein, A., and Infeld, L. *The Evolution of Physics.* New York: Simon & Schuster, 1938.
Frank, P. *Philosophy of Science.* New York: Prentice-Hall, 1957.
Hamming, R.W. "The Unreasonable Effectiveness of Mathematics." *American Mathematical Monthly* 87 (1980):81–90.
Hanson, N.R. *Patterns of Discovery.* New York: Cambridge University Press, 1958.
Hardy, G.H. "Mathematical Proof." *Mind* 38 (1928):1–25.

Hempel, C.G. "Geometry and Empirical Science." *American Mathematical Monthly* 52 (1945):7–17.

———. "On the Nature of Mathematical Truth." *American Mathematical Monthly* 52 (1945):543–556.

Jeans, Sir J. *The Mysterious Universe.* New York: Macmillan, 1930.

Kitcher, P. *The Nature of Mathematical Knowledge.* New York: Oxford University Press, 1983.

Körner, S. *The Philosophy of Mathematics.* London: Hutchinson University Library, 1960.

Lindsay, R.B. *The Nature of Physics.* Providence, R.I.: Brown University Press, 1968.

Peynson, L. "Relativity in Late Wilhelmian Germany: The Appeal to a Preestablished Harmony between Mathematics and Physics." *Archive for History of Exact Sciences* 27 (1982):137–155.

Poincaré, H. *The Foundations of Science.* Lancaster, Pa.: The Science Press, 1946. The book contains English translations of several of Poincaré's expository books originally published separately, namely, *Science and Hypothesis, The Value of Science,* and *Science and Method.*

———. *Last Thoughts.* New York: Dover Publications, 1963.

Randall, J.H., Jr. *The Making of the Modern Mind,* rev. ed. New York: Houghton-Mifflin Co., 1940.

Weyl, H. *Mind and Nature.* Philadelphia: The University of Pennsylvania Press, 1934.

———. *Philosophy of Mathematics and Natural Science.* Princeton, N.J.: Princeton University Press, 1949.

Wigner, E.P. "The Unreasonable Effectiveness of Mathematics in the Natural Sciences." *Communications on Pure and Applied Mathematics* 13 (1960):1–14.

CHAPTER XIII

Bohm, D. *Causality and Chance in Modern Physics.* London: Routledge & Kegan Paul, 1957.

Bunge, M. *Causality and Modern Science,* 3rd rev. ed. New York: Dover Publications, 1979.

———. *The Myth of Simplicity.* Englewood Cliffs, N.J.: Prentice-Hall, 1963.

Burtt, E.A. *The Metaphysical Foundation of Modern Science,* rev. ed. London: Routledge & Kegan Paul, 1932.

Cassirer, E. *Determinism and Indeterminism in Modern Physics.* New Haven, Conn.: Yale University Press, 1956.

Crombie, A.C. *Turning Points in Physics.* New York: North-Holland Publishing Co., 1959.

Frank, P. *Modern Science and Its Philosophy.* New York: George Braziller, 1941.

———. *Philosophy of Science.* Englewood Cliffs, N.J.: Prentice-Hall, 1957.

Guillemin, V. *The Story of Quantum Mechanics.* New York: Charles Scribner's Sons, 1968.

Harré, R. *The Philosophies of Science.* New York: Oxford University Press, 1972.

Heisenberg, W.K. *Physics and Philosophy.* New York: Harper & Brothers, 1958.

Lucas, J.R. *Space, Time and Causality.* New York: Oxford University Press, 1983.

Margenau, H. *The Nature of Physical Reality.* New York: McGraw-Hill, 1950.

Reichenbach, H. *Atom and Cosmos, the World of Modern Physics.* New York: George Braziller, 1957.

Russell, B. *A History of Western Philosophy.* New York: Simon & Schuster, 1945.

Schrödinger, E. *Science and the Human Temperament.* New York: W.W. Norton & Co., 1935.

Toulmin, S. *The Philosophy of Science.* London: Hutchinson University Library, 1953.

Weyl, H. *Philosophy of Mathematics and Natural Science,* Princeton, N.J.: Princeton University Press, 1949.

Index